the
TOILET PAPERS

Interior at Farallones Institute Rural Center composter. Low seat platform permits sitting or squatting.

the
TOILET PAPERS

Recycling Waste and Conserving Water

Sim Van der Ryn
Foreword by Wendell Berry

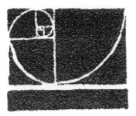

Ecological Design Press

Acknowledgments

To my friends who are slaying the kaka dragon with calm passion, common sense and good humor: Tim and Karin Winneberger, Peter Warshall, Orville Schell, John Poorbaugh, David Katz, Max Korschel, Saul Krimsly, Anon Forrest, Wade Rose, David Chadwick, Gib Cooper, Abby Rockefeller, Harold Leich, Bill and Helga Olkowski, Tom Javits, Sterling Bunnell, and to the many unseen friends who wrote in for "privy counsel."

My thanks also to Don Ryan for his fine illustrations; to Wayne McCall for his sensitive photography; and to the folks at Capra Press, especially Marcia Burtt and Noel Young.

—Sim Van der Ryn
"Captain Compost"
Sacramento, December 1977

Illustration Credits

Pages 18 and 19, Lawrence Wright, *Clean and Decent,* University of Toronto Press, 1967; page 21, J. Donkin, *Dry Sanitation vs. Water Carriage,* London, 1906; page 25, Alan Watts, *In My Own Way,* Random House; page 28, Collection Edwin Janss; pages 29 and 30, courtesy Labelle Prussin; page 31, Lawrence Wright, *Clean and Decent;* page 34, courtesy City of Los Angeles; page 36, *Sauvage* Magazine, Paris; pages 39 and 40, Clivus Multrum, USA, Cambridge, Mass.; page 42; Recreation Ecology Conservation of U.S., Milwaukee, Wis,; page 43, Blankenship Research Products, Dallas, Texas; page 44, Monogram Industries, Venice, Cal.; page 48, Shasta County Health Department; page 50, *Stop The Five Gallon Flush,* McGill University, Montreal, Canada; page 80, Farallones Institute, Berkeley, Cal.; page 85, Peter Warshall, *Septic Tank Practices;* page 93, courtesy Great Circle Associates, Kensington, Cal.; page 98, Farallones Institute, Berkeley; pages 107, 108, and 110, Solar Aquafarms, Inc., Encinitas, California; all other photographs by Wayne McCall.

Library of Congress Cataloging in Publication Data

Van der Ryn, Sim.
    The toilet papers: recycling waste and conserving water / Sim Van der Ryn.
        p.   cm.
    Originally published: Santa Barbara: Capra Press, 1978.
    Includes bibliographical references and index.
    ISBN 0-9644718-0-9
    1. Sewage—Purification—Biological treatment. 2. Land treatment of wastewater. 3. Compost. 4. Toilets. I. Title.
    TD755.V36   1995
    628.3—dc20

Ecological Design Press, Ten Libertyship Way, Suite 185, Sausalito, California 94965

Throughout this book you will find the word "waste" used to refer to those raw materials—feces and urine—your body passes on to make energy available to some other form of life. This is what you give back to the earth. The idea of waste, of something unusable, reveals an incomplete understanding of how things work.

Nature admits no waste. Nothing is left over; everything is joined in the spiral of life. Perhaps other cultures know this better than we, for they have no concept of, no word for, waste.

A sound man is good at salvage,
At seeing nothing lost.

—*Lao Tze, 500 B.C.*

CONTENTS

1. Notes on the History of Easing Thyself   17
2. Meet Your Wastes   33
3. Dry Toilets   37
4. How To Build Your Own Compost Privy   57
5. Household Composting   75
6. Greywater Systems   81
7. The Urban Sewer   99
    Epilogue   117
    Bibliography   121

FOREWORD

If I urinated and defecated into a pitcher of drinking water and then proceeded to quench my thirst from the pitcher, I would undoubtedly be considered crazy. If I invented an expensive technology to put my urine and feces into my drinking water, and then invented another expensive (and undependable) technology to make the same water fit to drink, I might be thought even crazier. It is not inconceivable that some psychiatrist would ask me knowingly why I wanted to mess up my drinking water in the first place.

The "sane" solution, very likely, would be to have me urinate and defecate into a flush toilet, from which the waste would be carried through an expensive sewerage works, which would supposedly treat it and pour it into the river—from which the town downstream would pump it, further purify it, and use it for drinking water.

Private madness, by the ratification of a lot of expense and engineering, thus becomes public sanity. This is permitted by our habitual disregard of consequences. We live by buying and selling the causes of every conceivable blight from cancer to famine to holocaust—and are continually astonished to find that these causes have their inevitable effects. As a society, we never look behind us at the generations that will follow us and at the impediments we are throwing in their way.

The importance of this little book is that it begins in the awareness of effects. It proposes to solve the sewage problem by doing away with its cause. This solution springs from an elementary insight: it is possible to quit putting our so-called bodily wastes where they don't belong (in the water) and to start putting them where they do belong (on the land). When waste is used, a liability becomes an asset, and the very concept of waste disappears.

All this, of course, is the commonest of common sense. And, of course, it will seem outlandish and revolutionary to the sanitation engineers, the public health officials and experts, and the manufacturers who are mining their dollars out of the "sewage problem," and who therefore have little interest in solving it.

Meanwhile, Sim Van der Ryn and the waste-users he speaks for will have the comfort of being right. They are working at the beginnings of an authentic sanity.

—WENDELL BERRY
Port Royal, Kentucky

PREFACE TO THE 1995 EDITION

It has been a number of years since my obsession with the issues presented in this little volume caused friends and neighbors to call me "Captain Compost." Since then, many more people have become aware of how precious pure water is and how important the nutrient cycle is to soil fertility.

Since this book was written, many composting privies have been built in varied climates. Home-built systems in rural California were studied and monitored as part of a state-funded study. I offer the following observations, based on talking to many compost privy owners over the years. It takes mindfulness and dedication to keep a composting privy working optimally over a long period. Most people don't do it. Turning the pile is a task few people turn to enthusiastically. I've found that most compost privies function more like storage vaults, safely storing human wastes until emptied and hot composted along with garden and other organic materials. For odor-free use, it's important to keep moisture down, using sawdust or other dry carbonaceous material.

Since the book was written, other options have come into use. Low flush toilets using from two quarts to a gallon-and-a-half per flush and engineered raised-bed leaching fields make it possible to accommodate standard septic systems in difficult soils. Health authorities continue to oppose composting privies.

The critique of conventional sewage treatment stands. If anything, there is increasing scientific confirmation of the environmental and health hazards posed by conventional systems. When *The Toilet Papers* was written, biological treatment and wetlands reclamation were in their infancy. In the next several generations, I expect that biology will win over concrete.

A new book on environmentally responsible small-scale home systems is needed. Until then, *The Toilet Papers* can be your guide in learning how to take greater responsibility for your place in the resource recycling web—the great continuing dance of life.

—SIM VAN DER RYN
Inverness, California

INTRODUCTION

Put yourself in the position of a future archaeologist sifting through the material remains of our culture some hundreds of years from now. What will he make of the curiously shaped ceramic bowl in each house, hooked up through miles of pipe to a central factory of tanks, stirrers, cookers and ponds, emptying into a river, lake, or ocean? "By early in the twentieth century urban earthlings had devised a highly ingenious food production system whereby algae were cultivated in large centralized farms and piped directly into a ceramic food receptacle in each home."

Our future archaeologist would need to be a genius to guess at the destructiveness and irrationality of present-day "sanitary engineering."

Mix one part excreta with one hundred parts clean water. Send the mixture through pipes to a central station where billions are spent in futile attempts to separate the two. Then dump the effluent, now poisoned with chemicals but still rich in nutrients, into the nearest body of water. The nutrients feed algae which soon use up all the oxygen in the water, eventually destroying all aquatic life that may have survived the chemical residues.

All this adds up to a strange balance sheet: the soil is starved for the natural benefits of human manure, garbage and organic materials that go down the toilet, the drain and to the dump. So agribusiness shoots it up with artificial fertilizers made largely from petroleum. These synthetics are not absorbed by the soil and leach out to pollute rivers and oceans. We each use eight to ten thousand gallons of fresh water to flush away material that could be returned to the earth to maintain its fertility. Our excreta—not wastes but misplaced resources—end up destroying food chains, food supply, and water quality in rivers and oceans.

Nations endure only as long as their topsoil. How did it come to pass that we devised such an enormously wasteful and expensive system to solve a simple problem? Excreta is one of few substances of material value we ever return to the earth. Our body waste is truly a resource out of place.

Sir Albert Howard suggests that a primary criterion for a successful culture is to realize a balanced relation between the processes of growth and the processes of decay. He notes that our society, which exclusively values growth and looks upon the processes and products of decay as waste, is radically out of balance.

The way we treat shit also reflects our attitude towards the body and its functions. The development of Mr. Crapper's water closet and urban sewer systems coincides with the ascendance of Victorian priggishness typified by clothing that disguised the body's form from head to foot. The gleaming white functional bathroom was perfected in the twenties—a period noted for its crusade against germs, those nasty creatures in the mouthwash ads. One wonders how the bacteria that sustain our lives ever survived the antiseptic hygiene age.

East and West developed very different attitudes and practices in relation to the human body and its processes. In China and Japan, "night soil" has been scrupulously collected for centuries to fertilize the fields. A nineteenth-century visitor to Japan tells us that in Hiroshima, in the renting of poorer tenement houses, if three persons occupied a room together the sewage paid the rent of one, and if five occupied the same room no rent was charged. Farmers vied with each other to build the most beautiful roadside privies in hopes of attracting the favors of travelers who needed to relieve themselves.

Rational disposal systems in the Orient grew out of the importance of excreta to agriculture. Carts traveled through the cities collecting the precious stuff and carrying it off to dung heaps where it decomposed. In the West no such practice existed. Chamber pots were emptied into the back yard or street. Some of the streets were designed so that gutters would carry off the filth during a rain. Most of the time, city streets were not pleasant places to be: it is easy to smell how shit got a bad name.

In nature, water carries off wastes, and excreta is just another nasty waste. Early sewerage systems emulated natural process. The open gutters, washed clean only by rain, were gradually put underground to minimize the appalling stench and mess. In the 1800s, it was discovered that many then-common epidemic diseases were transmitted through microorganisms in feces. But by then the psychological and technological die had been cast. The basically unsound practice of dumping excreta into any convenient body of water was rationalized. The flush toilet eliminated direct contact with excreta. The smell and mess were removed from city streets and put into underground pipes. Methods to treat sewage by settling out solids, adding chemicals to kill bacteria, and, more recently, aerating to speed decomposition, were invented.

We assume that by flushing and forgetting we are rid of the problem, when we have only compounded it by moving it to another place. Every tenderfoot camper knows not to shit upstream from camp, yet present urban culture provides us no alternative. It is estimated that a quarter of all urban sewage is dumped into the water. The rivers, bays, and oceans around half our urban areas are cesspools. The waste we seek so hard to ignore threatens to bury us.

the
TOILET PAPERS

Chapter 1.

Notes on the
History of Easing Thyself

When you get right down to it, the basics of life are sleeping, eating, working, loving—and relieving oneself. So it's not surprising that the beliefs, practices and technology surrounding elimination are central to any culture. In many rural and seminomadic societies the act is casual and unselfconscious. People relieve themselves in fields or bushes and nature does the rest, often with an assist from wandering dogs, pigs and chickens, who feed on fresh feces.

Moses, leader of a nomadic Semite tribe, said, according to the King James version of Deuteronomy: "Thou shalt have a place also without the camp . . . and it shall be when thou wilt ease thyself, thou shalt dig therewith, and shalt turn back and cover that which cometh from thee." In other words, find a place away from camp to ease thyself and bury it. Moslem doctrine prescribes strict procedures to limit contact with fecal material. Only the left hand can be used for cleansing after elimination; the right is used for eating.

Observant Hindus consider it ideal to pass a stool first thing in the morning. Strolling a "bow shot" away from the home, the Hindu carries a brass vessel filled with water to a secluded spot away from running water, public roads, or temples. New garments are not worn, conversation is avoided. The feet are washed before elimination and the anal region is cleansed with water afterwards. Ending the ritual is symbolized by rinsing the mouth eight times with water. Religious rituals such as this serve as a guidepost to hygiene, sometimes

17

reflected in class distinctions. The chore of the caste of Untouchables was to carry away nightsoil. Among Moslems and Jews, prohibitions against eating "unclean" meat, such as pork, probably stemmed from the observed habits of the animal. The value placed on running water in the Moslem and Hindu traditions has a similar hygienic basis.

I cannot always find clear physical reasons why some cultures disposed of their wastes in water while others used the land. South Sea islanders build their privys over the ocean. The western tradition of sewerage comes from the Romans who built both aqueducts for water supply and sewers leading to the Tiber. Perhaps the earliest known water-based fixture was the squat-type toilet used in the Palace of Knossos four thousand years ago—similar to those used all over the Mediterranean area today.

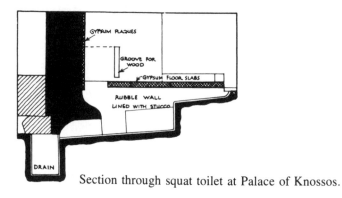

Section through squat toilet at Palace of Knossos.

Bathing was scorned in medieval Europe. In the cities, waste was dumped haphazardly in the streets. Nobility did somewhat better. Their castles were equipped with *garderobes,* simple privys built into castle walls and often emptying into a stenchy moat.

Sewer and water closet devices made their appearance in the mid-nineteenth century in Europe, although types of dry toilets remained in use for many years.

EARTH CLOSETS AND HONEY BUCKETS

The earth closet is simply a toilet seat with a bucket below to receive

18

the waste matter. After every use, fine dry soil is added or manually flushed from a container; the bucket is emptied periodically. The

Garderobes at Southwell Palace.

Perhaps when Shakespeare wrote of being "shat on from a great height" he had the medieval garderobe in mind.

Garderobes at Langley Castle.

advent of sewers in the 19th century, and health problems resulting from water borne treatment, sparked an interest among medical people and sanitarians in earth closets. They remain among the simplest and most effective devices, especially when the contents are composted aerobically with other organic materials (see Composting section).

A variation of the earth closet, the can privy, was used in many American cities until the 1930s. The can privy was simply a removable seat on a can which was periodically collected by mule-drawn "honey wagons."

The arrival of central sewerage into Western Europe was a technological event that coincided with the emerging egalitarian ideal. For centuries, class distinctions separated the odor, dirt and smell of the Unwashed Masses from those privileged to escape the ritual of emptying slops into the street. The flush toilet made one

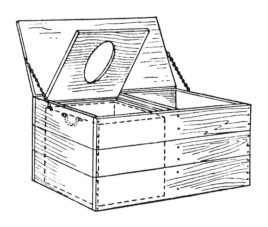

Practical earth closets found in a 1918 housewives' manual.

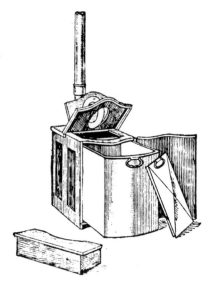

person's waste equal to another's in the great stream of sewage. In America, the census celebrated the spreading of democracy by noting the growing number of flush toilets. Today we are all equal in possessing this gleaming ceramic symbol.

As an egalitarian tool, it is used in utmost privacy, austerely and

An elaborate earth closet proposed by British architect J. Donkin in 1906. Note separate urinal emptying into earth-lined pan.

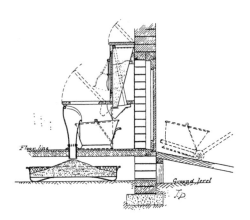

solemnly. I remember a professor of mine, a recently arrived Italian, who was most struck by the lack of conviviality surrounding the use of the American toilet. In his village, men joined each other at the village latrine for conversation and camaraderie at the first shared act of the day.

The can privy. Cans were periodically collected by "honey wagons."

WET AND DRY TOILETRY—EAST AND WEST

Asian cultures have traditionally valued human waste for agriculture.
In urban Japan, individual vaults are regularly pumped and material
taken to farm areas. In rural China, the pig has been rationalized into
the waste recycling system. A brick or stone lined latrine is connected
to the pig pen, so the pigs can eat freshly passed stools (although
recent barefoot doctor reports advise against this practice). Modern
China exhorts the people to "turn all waste into treasure" and to
consider the needs of the whole. Composting fecal waste is a
common practice. A British physician who spent fifteen years in
China reports:

> When the latrines are emptied the excrement, mixed with animal dung, is
> treated by the simple process of high-temperature composting. A mixture of
> equal parts of dung, earth, water, straw is piled into a specially prepared pit in
> which vertically placed sheaves of maize stalks provide ventilation. Gradually,
> the temperature of the compost rises high enough to kill the harmful worm eggs
> and make it safe for use as fertilizer. After high temperature composting, the
> manure becomes lighter in weight and easier to spread while its effectiveness as
> fertilizer is enhanced.

—Dr. Joshua S. Horn, *Away With All*
Pests: An English Surgeon in
People's China

23

In North Vietnam during the 60s, thousands of vault privys were built as part of a widespread program of rural sanitation. The concrete tanks consisted of two watertight compartments with a top squat plate. When one side is almost full, green leaves are added and the hole is sealed. After three months of decomposition, the material is removed for agricultural use. Meanwhile, the other side is used and the cycle begins again.

Still another sewage disposal system in the Manchurian city of Changchun is described:

> "A ten kilometer long sewage disposal canal has been built in Changchun. It carried off sewage water which used to flow into the Yitung River which cuts across the city. The sewage water (treated) drains off to irrigate farm land on the outskirts. This has helped to improve the city's environmental sanitation and increase production in grain and vegetable output. Now 52,000 tons of sewage water from the city's factories and living quarters drain off to irrigate 330 hectares of paddy field and 1200 hectares of other cropland. The people's communes, using sewage water which contains nitrogen and phosphorus to irrigate their farmland, save a total of about 3500 tons of chemical fertilizer a year."
>
> Rather than defining the problem in terms of "disposal," namely, how to get rid of treated or untreated effluent through an outfall pipe into a river or other receiving water, the Chinese stress "use." Sewage happens to be a valuable commodity while in America we are still installing primary and secondary treatment plants for sewage (namely, partially or wholly treating sewage, adding chlorine and getting rid of it). Meanwhile, the Chinese are moving toward ecologically sound and financially wise tertiary treatment plants which seek to re-use both water and solids. Nothing is disposed simply for the sake of disposing.

—Orville Schell, *All Waste Is Their Treasure*

Attitudes may be changing. Mr. Schell, a recent Chinese-speaking visitor who worked on an agricultural commune, told me that when he took pictures of the communal privy, people became upset, thinking he was focusing attention on their "backward" ways.

In his biography, philosopher Alan Watts writes of his bathroom in an early nineteenth century English parsonage:

> The bathroom was so abominable that I have made a drawing of it. It was constructed by a people, by a whole culture, which had never figured out delightful and amusing ways of handling such fundamentals of life as crapulation and bathing. Even an impecunious Japanese farmer has a reasonable bath where you can sit, soak and laugh with the rest of the family, but is kept separate from the craptorium, or *benjo*, which is operated on a system that, instead of wasting millions of gallons of water and requiring complex and

24

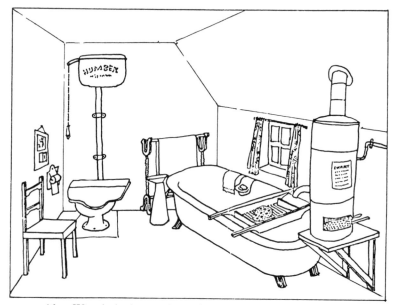

Alan Watts's bathroom in a nineteenth century parsonage.

ridiculous thrones, enables human excrements to be recycled and returned to the earth. The amount of shit we allow to flow out into the oceans is simply wasted manure—wasted, neglected, ignored because our eyes, noses and mouths go one way and our assholes go another.

—Alan Watts, *In My Own Way*

Edward Morse, a perceptive 19th century American traveler to Japan, wrote of Japanese country inns he visited:

It would be an affectation of false delicacy were no allusion to be made to the privy, which in the Japanese house often receives a share of the artistic workman's attention. . . . In the country the privy is usually a little box-like affair removed from the house, the entrance closed half way up by a swinging door. In the city house of the better class it is at one corner of the house, usually at the end of the verandah, and sometimes there are two at diagonal corners. The privy generally has two compartments,—the first one having a wooden or porcelain urinal; the latter form being called *asagaowa,* as it is supposed to resemble the flower of the morning glory,—the word literally meaning "morning face" (fig. C). The wooden ones are often filled with branches of spruce, which are frequently replenished. The inner compartment has a rectangular opening cut in the floor, and in the better class of privies this is provided with a cover having a long wooden handle. The wood-work about this opening is sometimes lacquered. Straw sandals or wooden clogs are often provided to be worn in this place.

The interior of these apartments is usually simple, though sometimes presenting marvels of cabinet-work. Much skill and taste are often displayed in the approaches and exterior finish of these places.

Figure A shows the interior of a common form of privy. Figure B illustrates the appearance of one in an inn at Hachi-ishi, near Nikko.

As one studies this sketch, made at an inn in a country village, let him in all justice recall similar conveniences in many of the country villages of Christendom!

In figure C is shown the privy of a merchant in Asakusa, Tokio. The door was a beautiful example of cabinet-work, with designs inlaid with wood of different colors. The interior of this place (figure D) was also beautifully finished and scrupulously clean.

Figure A—Interior of Privy.

Figure C—Privy Connected with a Merchant's House in Asakusa.

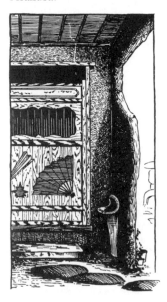

Figure D—Interior of a Privy in Asakusa.

The receptacle in the privy consists of a half of an oil barrel, or a large earthen vessel, sunk in the ground, with convenient access to it from the outside. This is emptied every few days by men who have their regular routes; and as an illustration of the value of this material for agricultural purposes, I was told that in Hiroshima in the renting of the poorer tenement houses, if three persons occupied a room together the sewage paid the rent of one, and if five occupied the same room no rent was charged! Indeed, the immense value and importance of this material is so great to the Japanese farmer, who depends entirely upon it for the enrichment of his soil, that in the country personal conveniences for travellers are always arranged by the side of the road, in the shape of buckets or half-barrels sunk in the ground.

Judging by our standards of modesty in regard to these matters there would

Figure B—Privy of Inn
in Hachi-ishi Village,
Nikko.

appear to be no evidence of delicacy among the Japanese respecting them; or, to be more just, perhaps I should say that there is among them no affectation of false modesty. Indeed, privacy in this matter would be impossible when it is considered that in cities—as in Tokio, for example—of nearly a million of inhabitants this material is carried off daily to the farms outside, the vessels in which it is conveyed being long cylindrical buckets borne by men and horses. If sensitive persons are offended by these conditions, they must admit that the secret of sewage disposal has been effectually solved by the Japanese for centuries, so that nothing goes to waste. And of equal importance, too, is it that the class of diseases which scourge our communities as a result of our ineffectual efforts in disposing of sewage, the Japanese happily know but little. In that country there are no deep vaults with long accumulations contaminating the ground, or underground pipes conducting sewage to shallow bays and inlets, there to fester and vitiate the air and spread sickness and death.

—Edward Morse, *Japanese Homes and Their Surroundings*

BALL COCK ON THE THAMES AND TOWERS ON THE PLAIN

Artist Claes Oldenburg brings the technology of water borne sewage literally out of the closet and makes it an active public monument, while the Mali villagers have for centuries created ecological monuments out of their unique privy towers.

For Oldenburg, the artist is someone who is involved with creating appropriate symbols for our time. Art, he believes, is still an area of metaphysics that can

28

explain and see the whole and has not relegated itself to analyzing and fragmenting. It is that area in which it is still possible to make broad universal statements about the nature of the human condition.

"The Thames Ball is a giant copper ball, based on the form of a toilet float, which is connected to the center of one of the bridges in the Thames. The ball rises and falls with the going out and coming in of the tide."

—from Claes Oldenburg, *Object into Monument,*
exhibition catalog, Pasadena Art Museum,
by Barbara Haskell

Without really thinking about it, the Western mind assumes that a waterborne waste disposal system is one of the requirements of an urban setting, and one of the symbols of an advanced society.

In actual fact, a waterborne disposal system can often prove disastrous to an economically viable community if cultivation depends heavily on soil fertilizer. A traditional solution to the problem of waste disposal in such a community is

The latrine shaft in Mali, showing the way one
city has solved the problem of sanitation without
a waterborne system of waste disposal.

29

found in Mali, in the city of Djenné, a tightly knit and densely populated city with a population of over 6,000. Here latrines are located on a second or third level terrace, and waste is accumulated in a well-reinforced pottery shaft. The shafts limit the danger of infection and disease, serving at the same time as excellent storage for the supply of nutrients.

—Labelle Prussin, *University of Michigan Research News,* Spring 1975

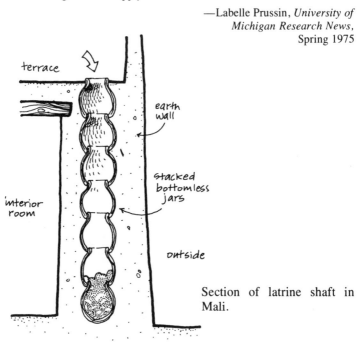

Section of latrine shaft in Mali.

TO SIT OR SQUAT?

The water closet was a British invention that introduced not only water to the toilet, but unhappily, sitting as well. For the first time people sat to move their bowels. The idea of defecating while on a piece of furniture fourteen inches off the floor seems to have originated with the portable thrones called "close stools" used by English and French kings, whose majesties were obviously threatened by squatting like the common folk.

A stool made in 1547 for the use of the kynges mageste was covered with black velvet and garnished with ribbons, fringes and 2,000 gilt nails. The seat and arms were covered with black cotton and fitted with straps, and may well have accompanied Henry VIII on his travels.

—Lawrence Wright, *Clean & Decent*

Royal Close Stool in Hampton Court, c. 1600.

The toilet may be beautiful but it is a design disaster as far as human physiology is concerned (although the makers of hemorrhoid and constipation remedies may not agree). Of course, there is something you can do if you have a conventional toilet: squat on top of the seat. You'd be surprised how many people do! From the squat toilets we've built, we've come to know the problems. It's best to have a grab bar to hang onto. In winter many layers of clothing get to be a problem, and women sometimes complain of difficulty in aiming their stream. But it is the most healthful way to go.

QUOTES ON THE VIRTUE OF SQUATTING
from Ken Kern, *The Owner Built Home*

The ideal posture for defecation is the squatting position, with the thighs flexed upon the abdomen. In this way the capacity of the abdominal cavity is greatly diminished and intra-abdominal pressure increased, thus encouraging the expulsion of the fecal mass. The modern toilet seat in many instances is too high even for some adults. The practice of having young children use adult toilet seats is to be deplored.

—Bekus, "Gastro-Enterology" p. 511

Man's natural attitude during defecation is a squatting one, such as may be observed amongst field workers or natives. Fashion, in the guise of the ordinary water closet, forbids the emptying of the lower bowel in the way Nature intended. . . . It is no overstatement to say that the adoption of the squatting attitude would in itself help in no small measure to remedy the greatest physical vice of the white race: constipation.

—Hornibrook, "The Culture of the Abdomen" p. 75

31

It should be mentioned in this connection that a very common cause for unsatisfactory results ... is improper height of the toilet seat. It is usually too high. An ideal seat would place the body in the position naturally assumed by man in primitive conditions. The seat should be low enough to bring the knee above the seat level.

—Williams, "Personal Hygiene Applied"
p. 374

The high toilet seat may prevent complete evacuation. The natural position for defecation, assumed by primitive races, is the squatting position. When the thighs are pressed against the abdominal muscles in this position, the pressure within the abdomen is greatly increased so that the rectum is more completely emptied. Our toilets are not constructed according to physiological requirements.

—Aaron, "Our Common Ailment" p. 66

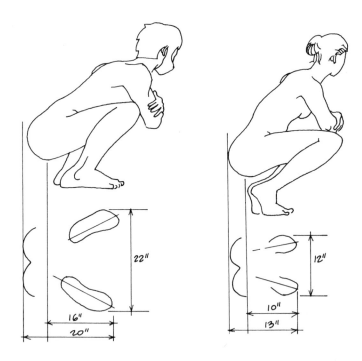

Normal squatting position for men and women.

32

Chapter 2.

Meet Your Wastes

The feces of an average healthy adult varies in size from 4 to 8 inches by 1 to 1½ inches and in weight from 100 to 200 grams. Both the consistency and odor as well as size and weight may vary considerably . . . the stool resulting from a basically vegetarian diet is generally larger, softer and less odorous . . . food residues account for only part of the total bulk of feces. The remainder is made up of bacteria, dead cells . . . feces consist of 65% water, 10 to 20 percent ash, 10 to 20 percent soluble substances and 5 to 10 percent nitrogen.

—Kira, *The Bathroom*

About 20 percent of the mass is a solid aggregate of live bacteria—more than 10 trillion each day—or more than the number of cells in your body. The intestinal tract is colonized by a great number of bacterial species, concentrated at the two extremes of the alimentary canal, the mouth and the large intestine, there are more than 100 different anaerobic species in a healthy stool, over 100 billion bacteria per gram.

—"Ecology of the Intestinal Tract"
Gerald T. Kersch,
Natural History, Nov. 1974

You pass, on the average, a quart of urine per day. The dry solids in urine—everything left when all water is removed—weigh about the same as the solids in shit. Urine is rich in nitrogen (15-10%) which means that you piss away 10 pounds of pure nitrogen each year.

"Man must do something with his wastes and the ocean is the logical place for them," was the subheading of an article "The

Disposal of Wastes in the Ocean," by Williard Bascom, that appeared in the August 1974 issue of *Scientific American*. A photo of the huge Hyperion Plant appeared with the following caption: "Sewage treatment plant serves the city of Los Angeles, discharging into the Pacific Ocean about 235 million gallons per day of primary treated effluent and 100 million gallons per day of secondary effluent. The discharge pipe is 12 feet in diameter and nearly five miles long. At the discharge end it is in 197 feet of water. The plant separately discharges sludge, consisting of about 1 percent solids, through a seven mile pipe to a depth of more than 300 feet. The sludge is discharged at the brink of a marine canyon."

That got me to thinking. The nutrients in all that effluent, much of it from flush toilets, if converted to fertilizer would be the equivalent of 200 tons of 7-14-12 fertilizer (7% nitrogen, 14% phosphorus, 12% potassium). Each ton of fertilizer when applied to soil would provide the nutrients to grow 25 tons of vegetables. Thus *each day* L.A.'s

The Hyperion Sewage Treatment Plant, Los Angeles.

34

waste provides the nutrients to grow 5,000 tons of vegetables, enough to provide everyone in Los Angeles with a pound or two of fresh produce daily.

As a footnote to this story, a friend sent me some clippings from the *Los Angeles Times,* 1971. It seems that nature lovers expressed horror over an L.A. County Department of Parks and Recreation project in which volunteer divers were sent into coastal waters to kill sea urchins. The sea urchins, thriving on the wastes dumped by the plant, had wildly proliferated, eating all the giant kelp which provides the habitat and food source for fish and most other coastal marine life. And so it goes once the circle is broken.

Chapter 3.

Dry Toilets

In the dry, or waterless, toilet, water is not used to dilute or transport human waste. The dry toilet converts waste to usable nutrients. Think of it this way: it takes forty tons of water a year to dilute a few hundred pounds of your wastes that when composted fit into several five gallon cans! The best way to have clean rivers, lakes and oceans is to avoid polluting them with waste water in the first place. Think of the savings in vast lakes of clean water that now go down the toilet, miles of pipe, acres of concrete, and billions of dollars of energy consuming public works that could be saved by using dry toilet systems.

Dry toilets have been designed to be used where water is not available in sufficient quantities, or where conventional systems may not work properly because of soil conditions or pollution potential. Dry toilet systems vary from the extremely simple—such as a hole in the ground—to the extremely complex, such as space age closed systems that require full-time technicians to operate. In recent times, dry toilet designs developed for a variety of reasons: effective low cost sanitation for the Third World and rural areas promoted by the World Health Organization and other groups; sanitary containment in areas where difficult site conditions or limited water supply demanded an alternative; or simply the desire shared by many people for a simpler, less wasteful and more ecologically sound way.

Most of this chapter focuses on the design of a home model you can build. There are, however, growing numbers of commercially available dry toilet systems. Most of them cost two to ten times as much as constructing a system yourself. All systems—commercial *and* owner built—require a greywater system to handle other household waste water, and must be approved by building and health departments for legal installation. Here is a quick review of some of the major types.

COMMERCIALLY AVAILABLE DRY TOILETS

Biological Toilets

Often called "humus" or "composting" toilets, these devices receive feces, urine, paper, and kitchen garbage into an aerated container where they are decomposed naturally by bacteria and other microorganisms. Since higher temperatures (approaching 160°F) produce more rapid decomposition and evaporation than lower temperatures, some units are equipped with an electric heating element to speed decomposition. Most moisture is evaporated off through the vent. Volume is greatly reduced—up to 90%. Small amounts of humus-like residue are left from the decomposition process which can be used as a soil enricher on non-edible plants.

Good aeration, moisture, and a temperature of at least 15°C (70°F) are necessary to achieve satisfactory decomposition. Excessive quantities of urine slow or stop the process. In such cases, dry, porous material like peat moss, sawdust, or ground leaves must be added to restore balance. Chemicals should never be added. All systems require a ventilation pipe—generally six inches in diameter or larger to carry away vapor and possible odor. In cold climates pipes should be insulated. Some systems use a mechanical fan for positive ventilation.

A layer of soil or garden compost helps the decomposition process get under way. Good soil is naturally rich in bacteria and molds. Porous humus may be added to absorb urine. Some toilets are equipped with a built-in rake to stir the waste and keep it from getting too dense. Biological toilets can breed flies, although a well functioning system is self-stabilizing. Controlling flies through gas-emitting strips or sprays should be introduced only as a last resort.

Twenty-one types of biological toilets were tested by the Microbiological Institute of the Agricultural College of Norway in order to discover whether disease causing bacteria and viruses survive a decomposition process. All toilets were contaminated with polio virus and salmonella. Samples of waste were extracted after various time lapses and tested for these pathogens. The results showed satisfactory sanitation in all toilets.

There are two favored designs for biological toilets. The first consists of a large box constructed of fiberglass or other impervious material with a bottom that slopes approximately 20 degrees. Seven feet of headroom is required for installation. A wide pipe connects directly to the toilet seat above with a separate chute for garbage. The angle and built-in air channels provide a draft upward through the waste mass, while allowing the decomposing waste mass to move slowly towards an access hatch. There are no movable parts.

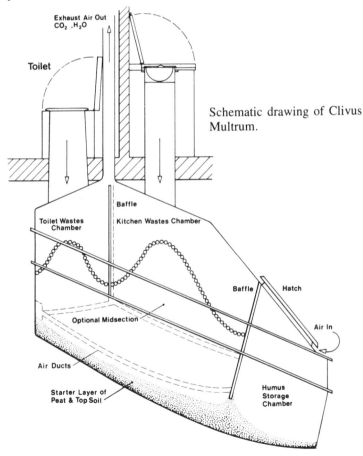

Schematic drawing of Clivus Multrum.

Decomposition occurs through low temperature moldering and composting, aerobic at the surface levels and anaerobic underneath. In normal household use, it takes several years to begin to produce finished humus, depending on climate and loading rates. Presently designed units provide for year-round use of up to six persons. Thousands of this type of unit are in use in Scandinavia. Acceptance by U.S. officials has been slow and the number of installations by 1978 is fewer than 10,000.

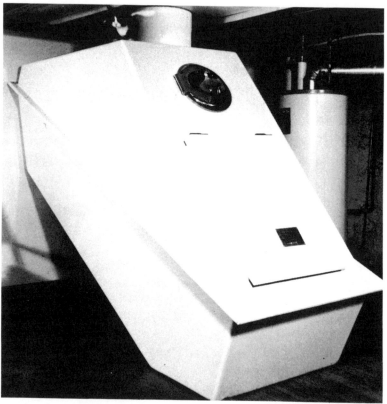

Installation of a Clivus Multrum in a basement.

In the second type of biological toilet, the seat is mounted directly over a container equipped with a heating element to promote evaporation and a constant temperature for decomposition as well as a fan for ventilation through a wall or ceiling vent. These mechanical

Topside of Clivus Multrum unit at Farallones Urban Center, Berkeley.

aids make it possible for the storage chamber to be much smaller than in the first type of humus toilet since decomposition and dehydration occur more rapidly. Units of Scandinavian design are used mostly for vacation cabins. There is no available data on year round use. The unit requires 110v electricity and is said to use 20 to 30 kilowatt-hours per month.

Incineration Toilets

These self-contained units incinerate toilet wastes; they have either electrical or gas connections. After use, the lid is closed while the device undergoes an approximately 15-minute long incineration cycle. The residual ash is to be disposed regularly, depending on use. Energy consumption rates are high. The obvious question is: why use one resource to destroy another?

Oil Flush Toilets

The oil flush toilet uses a mineral oil to transport wastes to a holding tank for storage. The fixture has the appearance of a standard toilet. The wastes must periodically be pumped out and hauled to a land disposal site. The material cannot be composted because it is coated with oil. The systems are expensive, complex and subject to

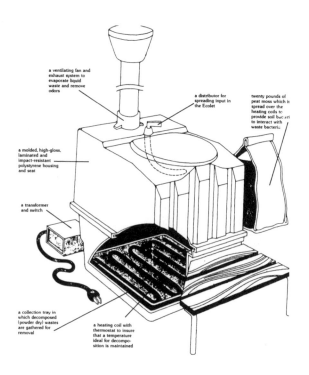

a ventilating fan and exhaust system to evaporate liquid waste and remove odors

a distributor for spreading input in the Ecolet

twenty pounds of peat moss which is spread over the heating coils to provide soil bacteri: to interact with waste bacteri:

a molded, high-gloss, laminated and impact-resistant polystyrene housing and seat

a transformer and switch

a collection tray in which decomposed (powder dry) wastes are gathered for removal

a heating coil with thermostat to insure that a temperature ideal for decomposition is maintained

Schematic drawing of Ecolet, biological toilet.

breakdown and costly maintenance. Oil flush toilets seem to be an example of how a mastery of complex technology combined with biological ignorance often complicates rather than simplifying a problem. Each unit costs almost as much as a compact car.

Chemical Toilets

Used widely throughout the United States and abroad, the chemical toilet is used mostly for temporary sanitation facilities at construction sites, athletic events, and farm camps. Small, portable units are manufactured for the camping market. Chemical toilets are constructed of fiberglass and are usually filled with a fragrant liquid and chemical with a high pH for preventing biological degradation between cleanings. Maintenance is via pumper truck which transports wastes to a sewage treatment plant or sanitary landfill.

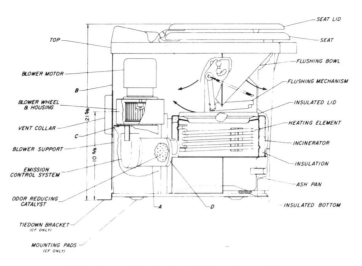

Side view of Incinolet, incinerating toilet.

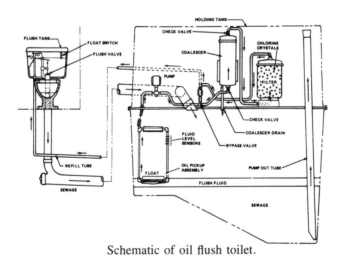

Schematic of oil flush toilet.

Low Water Use Toilets

There are many commercially available fixtures that reduce water use up to 90 percent. Standard tank type fixtures require five to seven gallons for each flush. Under a California law adopted in 1976, by the year 1979 all new fixtures must use no more than three gallons per flush. In England and Japan some fixtures have a two-position flush lever which dispenses less water to flush urine alone. Also in Japan, a fixture is available that designs the wash basin into the top of the toilet tank. The tank receives dirty wash water which is then used for flushing: a greywater toilet. Both compressed air and vacuum toilets are available for home use. The compressed air toilet flushes with a blast of water under pressure. About two quarts of water are required per flush. The distance of horizontal run to a septic or holding tank may be limited. Vacuum toilets have been used in apartment complexes in Scandinavia and elsewhere. Waste is sucked into the line upon flushing, along with a small amount of water. The material is held in a tank for pumping and disposal in landfill or sewage treatment plants.

HOME BUILT DRY TOILETS

Drum Privy

Homesteaders in Northern California in latter years have come up with a variation of the earth closet or can privy. This design uses a standard 55 gallon drum placed on a small scissors jack which is cranked up to provide a tight seal against the toilet seat above. The drum is vented, and sawdust or earth is added after each use. In an average family, it takes three to six months for the drum to fill up, at which time it is lowered, rolled out and sealed for composting in the drum, or picked up for composting at a central site by a local sanitation agency. Complete plans and guidelines can be obtained for $1.25 from Peter Warshall, Box 42, Elm Road, Bolinas, CA 94924.

The drum privy can be used to convert human waste to soil fertilizer by high temperature composting, to convert wastes to fertilizer by low-temperature composting, or to hold wastes within the drum until it can be transported to some other location for treatment. A drum privy described here is based on consultants' experience, especially that of Lon Hultgren, Peter Warshall, and Steve Matson.

One 55 gallon drum should be used for every two to three people. The drums will probably need rotation once every six to eight months. For a family of four, you would need two to four drums. Each drum should stand one year after filling to assure complete low-temperature composting.

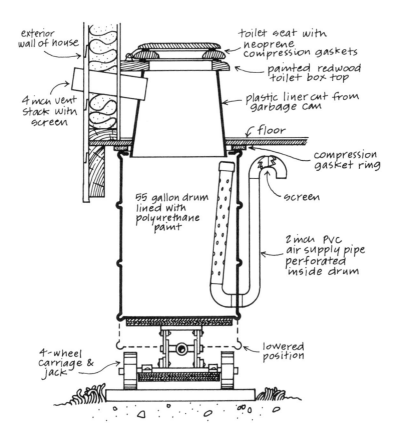

exterior wall of house

toilet seat with neoprene compression gaskets

painted redwood toilet box top

plastic liner cut from garbage can

4 inch vent stack with screen

floor

compression gasket ring

55 gallon drum lined with polyurethane paint

screen

2 inch PVC air supply pipe perforated inside drum

4-wheel carriage & jack

lowered position

Section through drum privy.

CONSTRUCTION, INSTALLATION AND DESIGN: If the drum is under a house or privy but exposed to sunlight, westerly orientation of the drum is desirable to take advantage of the sun's year-round warmth. The drum can be placed either above or below the ground.

Start with a 55 gallon drum. (For storage, drums can have a smaller volume.) Buy a drum with two inch threaded opening near the drum bottom. Choose drums without flaws. They should have a non-corrosive interior (polyurethane spray or asphalt emulsion). If the drum will be exposed to sunlight, it should have a flat black exterior. Drums should have lever-band lids.

Equip each drum with an aerator of durable, corrosion resistant materials. Aerators can be made of two inch perforated PVC or plastic air piping to fit the hole in the side of the drum. Screen the aerator, preferably with 22 mesh screen. Aerator tubing can be installed through the already existing hole at the side of the drum. Screw the aerator pipe into the air hole to seal the pipe to the drum.

Jack the drum as evenly as possible against the sub-floor of the privy, or bathroom. Do this with a dolly and jack or turnbuckles.

An optional drain can be placed at the very bottom of the drum by cutting a hole for a faucet. This drain will prevent puddling and the resulting odors by removing heavily contaminated liquid. A ¾" hole for a garden hose faucet is adequate.

A four to six inch PVC, plastic, or metal vent should extend from as high on the toilet chute as possible, through the seat to a point near the peak of the roof. If necessary, paint the vent stack flat black to increase air flow. Cap the stack to prevent rain from entering, and screen it (22 mesh) to exclude insects. If you live in a cold climate, insulate the stack to prevent condensation running back into the drum.

Special attention to weatherstripping and adequate seals is necessary on the toilet seat. Use a seat cover box with seals over a conventional toilet seat; neoprene gaskets work well on a standard toilet seat.

OPERATION: Eight to ten inches of dry sawdust or shavings should be placed on the drum floor. The drier the better, to absorb as much urine and liquid from the feces as possible. Shavings are better than sawdust because shavings allow more air circulation.

Two cups of sawdust or shavings should be added to the drum with each use. A supply of materials and an appropriate measuring device should be kept in a container next to the seat box or squat plate. Baking soda may be added for relief of any odors.

Use the drum until full. Then remove it and seal it with the vented lid. (If you have used your lid for a gasket, be sure to have an extra to seal the drum.) Install a fresh drum; store the full drum for pick-up, or compost the contents.

COMPOSTING:

Solar Composting: In hot sunny climates, the full drum can be stored in the sun. Each week the drum should be rolled vigorously. After three or four months of hot composting in the full sun, remove the lid and inspect the drum contents. If total composting appears to have occurred, the fertilizer can be used on ornamentals and trees. If not, reseal the drum and compost further.

Burial: Drum materials can be buried at almost any time after the drum is full. Contents should be buried in at least two to three feet of soil. Care should be taken when emptying the contents of the drum into the hole and replacing soil. This process should be done by people experienced in on-site waste management.

Outside-of-Drum Composting: Composting of the drum's contents outside the drum has the advantage of adjusting moisture and carbon-nitrogen ratio to obtain hotter piles. Trench composting is recommended over surface pile composting because there is less chance of human contact. Contents of the drum are placed in the center of a large (greater than one cubic yard) layered compost pile. Animal manure, water and plant material is added to get the highest temperature. Turn the pile every fourth day for a month, or until it no longer generates heat. At some point the pile should be at least 63°C (145°F). Between turnings the pile should be kept covered with a six-inch layer of finished compost, plant material or soil.

The Pit Privy

The best known dry toilet is, of course, the pit privy or old-fashioned outhouse. The butt of an older generation's rustic jokes, the pit privy is an effective sanitary device if properly located, constructed and maintained. Its drawbacks are that its use is limited to the most rural areas, it must be located some distance from the house, and it can be a source of unpleasant odors if there is too much moisture and decomposition becomes anaerobic. Avoid using lime or other chemicals which may slow or stop decomposition. A cup full of wood ash or dry soil can be added after each use.

Here are some typical health department recommendations for pit privy location, use and design.

SITE REQUIREMENTS: Pit privys should be installed only on

sites that fulfill the following requirements. (1) Pit privys should be installed only on sites where there is no evidence of fissured rock, and where slopes are not extreme. (2) Groundwater levels shall be at least five feet below the bottom of the pit during the wettest part of the year. (3) The privy must be located on the downhill side of the site, at least 150 feet away from any water supply.

DESIGN REQUIREMENTS: The pit privy should meet the follow-

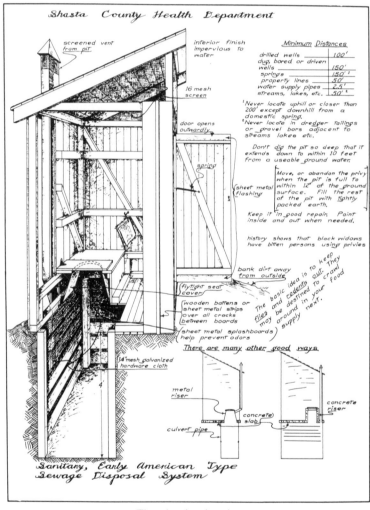

The classic pit privy.

ing design requirements. (1) Pits should be wood lined and have a minimum depth of five feet; 3½ × 3½ feet is a typical hole size. (2) Pits should be airtight except for the vent through the roof and the waste entry hole. (3) The area around the privy building and pit should be banked to divert surface waters away. The dripline of the privy roof should extend outside the diversion bank. (4) The waste entry hole should have a self-closing lid which, when closed, will effectively seal out insects or vermin. (5) A vent, extending above the roof line, should be provided into the seat box or through the privy floor. The vent should be designed to prevent rain from entering, should be effectively supported and rigged, and should be screened (16″ mesh) to keep out insects. (6) The privy building should have a supporting perimeter sill of concrete, six inch minimum cross section reinforced (or material of equivalent durability), a self-closing door, screened ventilation, and should be effectively sealed against insects and vermin. (7) If the pit privy is constructed on a slope or hill, a trench lined with gravel, rock or other durable material should be installed on the uphill side.

USE: Pit privys should be maintained in a sanitary condition. Keep door and seat closed when not in use. Wastes other than human feces, urine, toilet paper, and minor quantities of organic materials added for the purpose of odor control or pile digestion, should not be disposed of in pit privys. (A scoop of wood ash after every use reduces odor.) When waste fills the pit to a height of 18 inches below grade, the pit should be emptied or abandoned.

The Maine Tank/Clivus-Minimus

Plans for home built variations of commercially available humus toilets have been developed. They use the principles of a sloping box built out of poured-in-place concrete or concrete block with built-in air channels and roof vent. Their size can be varied to fit existing space; they can be built into existing cellars or into a new excavation alongside the house. Too few installations have been built, to my knowledge, to report on their operation.

The Farallones Compost Privy

I first developed this design based on experience with Swedish humus toilets, pit privys, and garden composting. The idea was to develop a low cost, owner-built dry toilet that could be designed into new or existing houses in the city or country—a refined indoor outhouse. The system would be easy to use and maintain, secure from insects and vermin, odor-free and sanitary, and would produce a usable humus end product. It would be comfortable to anyone familiar with garden composting.

The first prototype was built four years ago. Since then, through its publication in a Farallones Institute Technical Bulletin, many more

Design for owner-built Clivus composter.

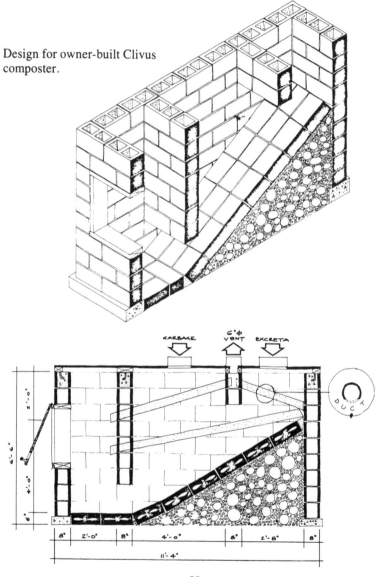

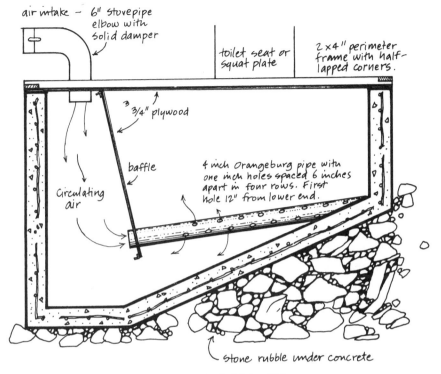

The home-built Maine Privy.

have been built throughout the country. Its design and management have evolved. I am grateful to Max Kroschel, whose careful research has resulted in improvements in its design and operation, and to the many owner-builders who have shared their experience with us. See the next chapter for how to build and manage this simple composting system.

The compost privy is designed to safely decompose human wastes without the use of water or plumbing. Household greywater from baths, kitchen and laundry, go into the sewer, septic tank or greywater systems described in this book. The volume of water used is reduced by almost half, pollution is reduced, and the processing and reuse of remaining wastewaters is simplified. The system is easily maintained to be odor- and nuisance-free by the organically minded householder. You are managing a complex biological system with no moving parts. The compost privy can be incorporated into a

51

Top side of a home-built sloping tank composter. Note foam
weatherstripping on seat cover to provide fly-tight seal.

new or existing one-story house in country, suburb, or city. With local approval, you can use the compost toilet where sewer hook-ups or septic systems are not possible. Besides saving water—up to 50,000 gallons per year in the average household—you will be able to return a small amount of your composted body waste to the soil each year in the form of dry, odorless humus.

The composting privy is designed to be built by the weekend builder who can pour a concrete slab, stack concrete blocks, and do simple carpentry. The materials, not including the toilet enclosure which you build to suit yourself, cost $200 or less (1978). The system consists of a two-compartment, above-ground, concrete block container, four feet wide, eight feet long and four feet high. It is best located on an outside wall, since access for turning and removal of material is required on one long side. The front of the three-sided block box is fitted with a removable plywood access panel with screened air inlets. A 12″ square vent stack must go a foot higher than the roof to allow venting of moisture and possible odors or gases.

A concrete curb keeps moisture from seeping out. A removable screened panel behind the access panel provides additional aeration and keeps material from falling out when the access panel is removed.

One compartment is used to receive fresh material. A conventional toilet seat or squat plate is located directly above its center. The receiving bin is lined with a layer of loose, dry material to absorb moisture and to provide aeration and microorganisms to begin decomposition. After each use a coffee canful of dry carbonaceous material —a mixture of peat moss, leaf mold, dry chopped grass clippings, sawdust, or shredded dry leaves—is added. A bin in the toilet room holds additive materials. The second chamber is for additional aging of the material. In normal family use, it takes about six months for one side to fill, then that bin is shifted to the other side for six more months of slow composting. After a year, the material should be fully decomposed and safe to fertilize non-edible plants.

Decomposition is produced by aerobic bacteria that require oxygen to multiply. The process starts without the addition of any substance other than human waste and other organic materials. Kitchen garbage is not added. Aeration is maintained by positive ventilation through intake vents and a vent stack as well as infrequent mechanical aeration with a manure fork. Cold outside temperatures slow down the process.

Guest composting privy at the Tassajara Zen Mountain Center. Excreta falls directly into a garden cart inside the holding chamber for recomposting with garden wastes at a separate location.

Interior of guest composting privy at Tassajara. Toilet compartments are to left and right.

Toilet compartment at Tassajara. Urine is collected separately. Note scoop for adding dry material.

Chapter 4.

How To Build Your Own
Compost Privy

COMPOSTING PRINCIPLES

Going into a strange kitchen I often have the urge to check out the garbage pail. Finding it filled with fruit peelings, spoiled leftovers, yesterday's newspaper, beer cans, glass jars, steak bones and plastic wrapping brings me to find my hosts and deliver a composting lecture. It was not always so. Ten years ago as I slurped the day's food garbage down the kitchen sink drain and into the disposal, a friend leaned over, "Don't you know that's all good stuff, you should be composting?" A bit peeved, and not knowing what compost was anyway, I told him to mind his own business in my kitchen, but a few weeks later he taught me to build my first compost pile.

Composting is easier to do than to describe, and, like lovemaking, magic when you do it well.

The ritual of turning inedible kitchen garbage, garden debris and other organic wastes into a finely decomposed material—humus—is a continual celebration of the wheel of life, connecting us and the fertility of the soil. Language recognizes the connection: *humanus* and *humus* are Latin for man and earth.

Many cultures have intuitively understood the cycle connecting decaying organic matter and soil fertility, but not until this century have safe, fast, reliable methods been developed through scientific experimentation and research. Composting organic material containing fecal matter requires extra care because feces may carry pathogenic or disease-carrying organisms. The principles described in this section apply to all composting and have been adapted to the design and management of the compost privy. (See Chapter 5 for methods of composting simple household and garden material.)

Composting is a managed process in which bacteria, fungi, and other organisms break down complex organic materials into simpler compounds, making their components available for use by plants once again. The finished, well decomposed organic matter, or humus, is a fine, dark earthy smelling material similar to what you find naturally under the leaf mat on a forest floor. The term compost comes from the Latin *compost* meaning "to bring together." The creatures that do the work include useful bacteria and other microorganisms not visible to the eye which are everywhere in our environment, as well as creatures you can see (macroorganisms)—detritivores feeding on decaying material, such as earthworms, grubs, insects and nematodes.

Two kinds of decomposition occur in nature. Aerobic decomposition occurs, for example, on the forest floor. Dead leaves, animal remains, feces and other materials are stirred and broken up by the passage of animals and insects. Bacteria that live in the presence of

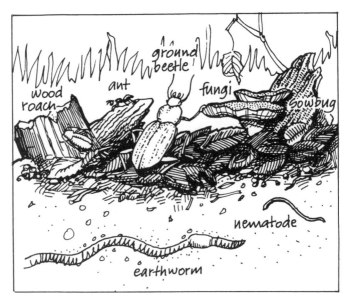

oxygen process the material through a series of chemical changes which reduce its mass to about one-twentieth of its initial volume. The results of the process are a nitrogen-rich, earthy humus, and carbon dioxide—both necessary to plant life. But in nature, the process of building topsoil through aerobic decomposition is extremely slow. It takes hundreds of years to build an inch of topsoil.

Anaerobic fermentation, in contrast to aerobic decomposition, is produced through the action of bacteria that live without oxygen, as in a swamp, marsh, or manure pile. Dead organic material goes through a series of chemical changes to produce humus, nitrogen, carbon dioxide and gassy by-products (including methane) that give the anaerobic process its distinctively unpleasant odor.

COMPOSTING YOUR WASTES

Because it is rapid, odorless and produces higher temperatures, aerobic decomposition is the process we want to encourage in the compost privy. To maintain the privy as an aerobic composter, consider the following factors:

Aeration: Aerobic bacteria live only in the presence of oxygen. Good aeration is insured by starting the pile with a layer of loose and dry material. Additional aeration is supplied by the airflow through intake vents and up the vent stack, and by turning as necessary. Frequent turning speeds decomposition and raises temperatures.

Moisture Content: The ideal aerobic compost pile is moist but not wet, and is fluffy and loose, not dense and matted. Since feces are 65-80 percent moisture and six to eight percent nitrogen, they must be balanced with three or four parts of a fine dry material. Moisture will need to be adjusted depending on climate and location.

Carbon/Nitrogen Ratio: Organic material contains varying amounts of carbon and nitrogen. Feces contain about six percent nitrogen, urine contains 15-18 percent. Since the optimum ratio for aerobic decomposition is 30 parts carbon to each part nitrogen, feces should not comprise more than a quarter of the pile.

Size and Temperature of the Pile: The optimum size of an aerobic compost pile is a cubic yard. Smaller piles don't provide enough insulating mass to hold the heat (the temperature at the center of a hotly decomposing aerobic pile reaches 160°F). Maintaining temperature requires supplying oxygen and fresh material. A common technique is to build the pile around a removable six to twelve inch diameter pipe, which will provide constant aeration.

BUILDING YOUR OWN COMPOST PRIVY

Constructing your own compost privy is not difficult. If you've ever poured a little concrete, laid concrete blocks and done simple carpentry you should have no trouble at all. Just simple handtools are required. I've talked to more than one woman who built the privy as her introduction to construction—they all did a fine job. Be sure to review the materials list and construction sequence before you start, though. The design has been refined after hearing from many folks who have built their own units.

Materials

This list may vary slightly depending on local conditions and details of your privy. It does not include materials for the enclosure, which you build to suit your needs.

Concrete for Pouring Slab and Filling Corner Blocks: Two-thirds cubic yard or twenty 90-pound sacks of redi-mix, or for home mix use six sacks of cement, six sacks of ¾″ gravel, five sacks of sand and five gallons of water. This makes one cubic yard.

Concrete Block: There are two types, standard and Fastblock, both eight inches wide. Fastblock is easier and quicker to lay up.

- 40 stretcher blocks
- 20 corner blocks
- 10 half blocks
- 4 90-pound sacks of redi-mix mortar, or one sack masonry cement with three parts sand

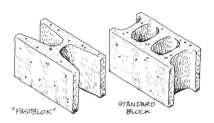

Hardware and Miscellaneous:
- 8 10″ foundation bolts (nuts and washers) to attach plates
- reinforcing rod and mesh (optional)
- 8⅜×7″ carriage bolts with wing nuts to fasten access panels
- 6 molly anchors and lag screws to fasten to runner for removable frames

- 2 square feet fine insect screening for vents
- 3 gallons asphalt emulsion to paint and seal inside
- caulking to seal cracks
- weatherstripping to seal access panel
- paint for access panel
- chicken wire for removable frames
- 8d and 16d galvanized nails
  Lumber:
- 3 2×6×8' redwood for plates and runner for removable frame
- 12 2×4×8' for framing top and access panel
- 2 sheets 4×8½ exterior plywood AC or AD grade for top and access panel
- Enough 2×4s and 1×2s to make slab form and stakes

Construction Sequence

Prepare to Pour Slab: Clear level area eight by ten feet. Make a form out of 2×4s with stakes at the corners and in the middle of the long side. Make sure form is level, square and firmly staked.

Pour Slab and Curb: Concrete ingredients can be hand mixed in wheelbarrow or trough. Use a large hoe or flat shovel and make sure ingredients are thoroughly mixed. Start placing concrete in the corner and spread with a shovel. When you are up to level, place a board across the forms and move it back and forth in a sawing motion to remove excess concrete and fill in low spots. Form and pour curb when slab is level and smooth but still wet. Let it cure overnight.

Lay Block Walls and Plates: Use garden trough and mix mortar in small batches, enough to use in one hour. Proper consistency will slide off trowel but won't sag. Lay first course in bed of mortar one half inch thick. Push blocks firmly in place. Check each block with a long level and tap firmly into place with trowel handle or hammer handle. Horizontal mortar joints should be ⅜" thick. Fill cores at anchor bolt locations with concrete and place anchor bolts with tops 3" above top of block. Punch or drill holes in front of block walls to receive carriage bolts that hold access panels. Make sure bolts are well positioned and grouted solidly. Lay a thin bed of mortar, set top plates and level. Paint inside of chambers with asphalt emulsion.

Framing: Frame floor and vent and assemble panels and removable frames. Use waterproof grade plywood and seal inside surfaces since moisture will condense on them. Attach runners for removable frames. Cut back top of outside runner so frames can slide forward for

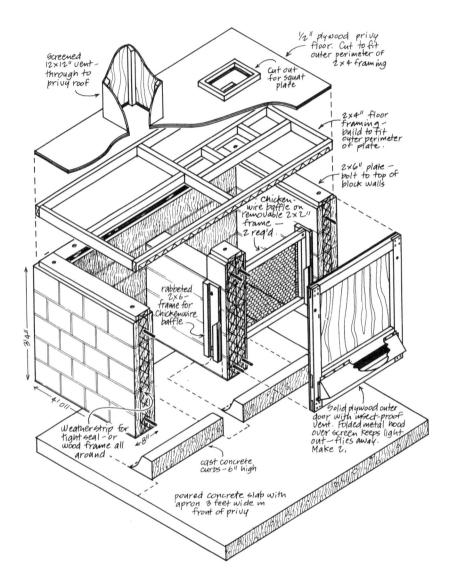

Screened
12×12" vent -
through to
privy roof

½" plywood privy
floor. Cut to fit
outer perimeter of
2×4 framing

Cut out
for squat
plate

2×4" floor
framing -
build to fit
outer perimeter
of plate.

2×6" plate -
bolt to top of
block walls

chicken-
wire baffle on
removable 2×2"
frame -
2 req'd.

rabbeted
2×6 -
frame for
chickenwire
baffle

34"

4'-0"

Weatherstrip for
tight seal - or
wood frame all
around

8"

cast concrete
curbs - 6" high

Solid plywood outer
door with insect-proof
vent. Folded metal hood
over screen keeps light
out - flies away.
Make 2.

poured concrete slab with
apron 3 feet wide in
front of privy

Exploded View—Farallones Compost Privy

62

easy removal. Cut out hole for squat plate or build box for conventional seat. If you use a squat plate, a grab bar two feet off the floor in front of the plate will be appreciated. Standard toilet seats do not make a tight enough fit to make a positive seal.

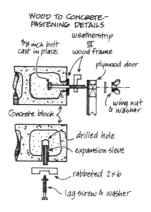

Sealing: Check fit on access panel. Tight seal is important as flies can enter through even an eighth-inch crack. Seal with weatherstripping or use a rabbetted jamb detail. Caulk and seal any other cracks. A sheet metal baffle over air vents will discourage flies.

Enclosure: Build the toilet room enclosure according to your own needs. Make sure all openings are screened, including the top of vent.

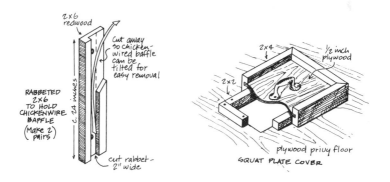

USE AND MANAGEMENT OF
YOUR COMPOST PRIVY

Start Up

Line the bottom of the collecting chamber with a six inch layer of sawdust and well decomposed compost or leaf mulch from the garden topped with six inches of chopped straw or dried grass. The compost provides a community of microorganisms to get things started, in addition to the bacteria in fecal matter. Sawdust is good for absorbing moisture and needs to be added only if you plan to pee into the privy. Straw provides some initial aeration.

Information Sheet

Post a sheet of basic information for guests on how to use the privy. Ours has a sign saying, "add one full can of material after each use." Replace lid after every use. Some people post a log including when the pile was started, turned, emptied, as well as notes of any problems. In privys receiving heavy use this information may be valuable.

Additives

Don't add metal, cans, plastic, boxes, unshredded paper, toxic liquids, chemicals, and particularly any chemical sanitary agents such as lime, lye, etc. Household garbage should be added sparingly or not at all since it tends to smell and attract fruit flies and house flies. Toilet paper, of course, is fine. Keep a storage bin, and a one pound scoop or can, filled with materials to add after use. The materials can be a mixture of any dry, relatively fine material with carbon or fiber content, dried lawn clippings, crushed leaves or mulch, wood ash, peat moss, grain hulls, chopped straw, whatever is easily available to you. If you have a shredder or grinder, materials can be run through it. Redwood or cedar sawdust, though good for odor control, should be used sparingly, since they don't compost well.

Urine

The privy accepts urine, although the added nitrogen and moisture may mean more frequent turning. Many people collect it separately in a bucket or five gallon can with a funnel. Dilute it five to one with water, add a little dolomite or gypsum to neutralize acidity and use it as a plant tea in the garden. Stale urine smells; use it up frequently. A year's worth of your urine adds up to 10-15 pounds of pure nitrogen.

The first composting privy I designed, built by Ken Sawyer and David Chadwick at Green Gulch Farm in 1974. The center box holds sawdust. Not shown is a regular height portable seat for visiting mothers-in-law. The shoji slides open to a beautiful view of soft hills.

Temperature

Monitoring a number of privys over recent years indicates that conditions required for high temperature aerobic composting are seldom present in the collection chamber. Rapid decomposition and temperatures of 160°F (68°C) require that materials be assembled all at once, that the pile be frequently turned, well layered, and that its mass approach a cubic yard. You can expect temperature of

A composting privy added onto student living quarters at the Tassajara Zen Mountain Center. All you need is an outside wall and a few steps up to the throne.

100-140°F. Heat falls off rapidly from the center of the pile outward. Because of the long retention time, high temperatures are not essential for proper management.

Turning

Family privys receiving normal use of four to six people may require little or no turning. The pile will build up slowly, and if there are no odor, fly or moisture problems, the chamber need be opened only for emptying.

Odor or the puddling of moisture can be corrected through turning and adding dry materials. Flies are attracted by odor and through openings. Turning and covering with dry material should correct it. Sticky fly paper (but not pest strips) can be placed in the chamber. For turning, use a close-tined manure fork reserved for privy use only. It should be stored inside the storage chamber or in a bucket of clean sand. During fly season, turn early in the morning before flies are active. Leave the concrete apron clean of any debris and replace the access panel as soon as you are done.

Storage

Periodically, or when the collection chamber is full, transfer material to the storage chamber for additional composting. Fresh materials should remain in the storage chamber for a minimum of six months before use.

When materials are transferred to the storage chamber, take care to build a pile that will heat up. Materials from the privy storage chamber should comprise no more than half the volume of the pile and should be layered with straw, garbage, garden waste, and sprinkled with water if necessary. Keep the fecal material to the inside of the pile, as temperature drops rapidly near the edges. In batches monitored at Farallones Institute, although the center of compost piles reached 60°C, the outside was only 40°C.

The plans in this book are for a normal family capacity; greater volumes will require additional storage space.

J.B. turns his compost every several months and adds it to his regular garden compost pile. The removable access panel matches exterior wall.

Volume

Human waste per person per day averages a half pound of feces (moist weight) plus one quart of urine. A yearly average is about 180 pounds of feces and 90 gallons of urine. At 11 pounds a gallon and seven gallons a cubic foot, this equals less than three cubic feet of feces and about 12 cubic feet of urine. Evaporation and decomposition reduce this raw wet volume to almost one-twentieth its original volume, or about one cubic foot per person per year. Government sources say to size a privy at two cubic feet per person per year. Figuring a volume of other organic waste five times that of human waste, two $3' \times 3' \times 3'$ compartments would serve a family of five.

Use of the Finished Compost

After six months to a year in storage with no fresh material added, your pile should be fully decomposed, free of pathogenic organisms,

Interior of compost privy added to a neighbor's existing house. Note bamboo grab bar and heavy duty removable cover.

and ready for use. As a soil amendment, health authorities may require test data as proof of safety, or burial in trenches with twelve inches of earth cover. Do not use finished compost on edible root or leaf crops. Save it for ornamentals, fruit trees, and other crops that do not come into contact with the soil.

Pathogens

The main purpose for sanitary procedures is to eliminate pathways to new hosts for disease-producing organisms that live in human intestines. Composting is designed to destroy these organisms. Pathogenic viruses, bacteria, and parasitic worms can, under the right conditions, cause a variety of diseases such as hepatitis (virus); and dysentery, typhoid, and cholera (bacteria and protozoa). In the U.S., the incidence of fecal borne diseases is extremely low. Few people carry the pathogenic organisms in their intestines and they are not, therefore, present in feces. Traditionally, these diseases are caused by the immediate pollution of drinking water by fecal matter, or by the use of infectious raw sewage as fertilizer on edible plants.

Besides water, which is the main route, other pathways for transmission of pathogens from infected materials are by vectors such as flies or rodents, or by direct contact with infected material by hand or mouth. Construction and management following the instructions in this section should provide a safe sanitary barrier.

If conditions are optimal, only a very few pathogenic organisms need be present to cause disease. Pathogenic organisms, however, coexist and compete with harmless microorganisms whose communities have astronomical numbers of cells.

> With a bacterial population in excess of 100 billion per gram of stool, organisms present in concentrations of only 10 million per gram would, on the basis of random chance, be isolated only once in 10,000 times.
>
> —"Ecology of the Intestinal Tract"
> Gerald T. Kersch, *Natural History,*
> Nov. 1974.

In other words, one might have to study 10,000 randomly isolated colonies from the intestinal contents to find a single cell of the species under consideration. Since sampling pathogens is difficult, the presence of the most common and harmless intestinal bacteria, *Escherichia coli,* is used as an indicator of live fecal material. The coliform count is a standard measure of fecal contamination, but is not a firm indicator that pathogens are present.

As interest in the use of sewage sludge and compost containing human wastes has grown, a number of studies have been done on the rates of pathogen survival, both in the laboratory and the field. Temperatures between •120-140°F (55-60°C) killed all common pathogens in an hour; it is hard to insure high uniform temperatures in all parts of the pile, however. Only sterilization by heat of all material is an absolute answer, but varied research evidence indicates that time and naturally antibiotic conditions outside the human intestine are sufficient to kill all common pathogens in a matter of months. Durable construction of the privy, limiting access by flies and other vectors to raw material; minimal direct contact with raw material; six months' aging; regular twice-a-year microbiological sampling; and use of material on non-edible plants only are adequate health safeguards.

Flies and fly breeding remain the greatest potential health nuisance. Flies are attracted by odor and moisture and a good aerobic compost pile should not attract flies. Keep the entrance to the privy well screened. Check the side edges of the access panel to make sure the flies cannot get in. During the warm months, turn the pile in the early morning when fly activity is low. Don't leave the access panel off any longer than necessary.

Health officials may be concerned that the privy is durably and adequately constructed to insure fly tightness. They want to be sure the management procedure minimizes handling and possible con-tamination and that potential pathogens will not be present in the end product, compost.

COMMONLY ASKED QUESTIONS

Will a privy function properly in cold climate or at high altitudes?

Most of our experience with composting privys has been in mild climates such as California with light nighttime frosts but no severe winter conditions. However, we are familiar with one remote location high in the mountains with occasional sub-freezing tempera-tures for weeks at a time that has had a number of communal size privys in operation for several years now. We also have reports of a privy being used in northern Kentucky for the last five years or so. They experience freezing winters with little adverse effect except slowing of decomposition. Neither owner reports any adverse effects from the cold temperatures or the altitude. The compost piles still heat up but decomposition is considerably slower on the outer edges of the piles adjacent to the cinderblock walls. The concrete floor and block walls act as a heat sink, robbing heat from the pile. To reduce heat loss through the foundation slab, styrofoam sheets can be used to insulate it. A 4'×8' sheet two inches thick can be placed in the formwork flat on the ground and the slab cast on top of it. Another insulator is vermiculite (expanded mica) commonly used as a potting compound and propagation medium. It can be substituted for gravel to make a lightweight concrete and cast in a three inch layer below the slab.

The walls of the vault can be made of some durable rot-resistant wood such as redwood or cedar. Two-by-six tongue and groove decking could make a vector-proof compartment. Cast the bolts for the mudsill into the slab. Grout the mudsill to the slab for a good seal.

Use an architectural grade caulk between the sides and the mudsill. In extremely cold climates the walls of the vault can be insulated with foam sheet or could be built up stud walls insulated with glass wool and sheathed with exterior or marine plywood protected with several coats of epoxy paint. If blocks are used, the standard cement block with air space inside, laid individually in mortar, would conduct less heat than speed blocks laid dry and infilled with concrete.

Can privys be used on two-story structures?

Composting toilets are generally unsuited to second story (or higher) installations. Feces and urine tend to adhere to the deep vertical walls of the shaft and eventually cause hard to solve odor problems. Updrafts can also be a problem, bringing odors into the toilet area and even blowing sawdust back up as it is being added after use. If a second story unit is used, the shaft should be tapered with the top narrower and getting wider as it goes down to minimize contact with fecal matter falling down the shaft.

Why doesn't the pile heat up?

A long probe thermometer available from instrument suppliers is a handy way to monitor the action of your compost in storage. If the piles do not heat up sufficiently (above 55°C for several days), you may wish to recompost the material in the center of a hot garden compost after the six-month aging period. You may also need to turn the pile more frequently before placing it in storage.

To facilitate adequate breathing of the compost pile in storage we make a tube-like cage out of half-inch wire mesh and stand it vertically in the center of the storage bin. Construct the compost pile around it, forming a ventilation chimney.

How much urine can the compost take? What do you do with an excess of urine?

You will know when you've got too much urine (nitrogenous material) as there will be an unmistakable ammonia odor. Also the pile will appear too moist. Add sawdust or peat moss. Excess urine can be collected in separate cans used as urinals, then diluted with water (one part urine to five parts water) and disposed of under ornamentals or fruit trees.

Can I go on vacation and let my privy sit for extended periods of time?

Sure. It will keep on composting, only much more slowly.

What do I do with my greywater?

See the chapter on greywater.

SANITATION, BUILDING AND HEALTH CODES

The Uniform Building Code definition of a habitable dwelling requires running water and a flush toilet. Plumbing design must meet the requirements of the Plumbing Code. If no public sewer is available, regulations generally require you to obtain an on-site sewage disposal permit from the local health department. Here is where many people have come to grief in trying to incorporate into their new or existing homes dry and composting toilets that do not use water.

Health Departments generally consider one type of on-site disposal system acceptable: the septic tank/leaching field combination. The indoor flush toilet was used for many years as the index of rural housing improvement and the septic tank/leaching field is a relatively simple, safe and maintenance-free solution and vastly preferable to central sewerage in non-urban areas.

In trying to get permits for composting toilets and the like, many people have run into a "Catch-22" situation. Health departments say they can't approve any system that doesn't use the flush toilet because it would violate the Building Code, and building departments say they can't consider alternatives until the health department approves them. Health departments want data before approval, but there is no way to get data without first setting the system in place. In some localities alternative systems have been approved as long as the homeowner agrees to provide a standard system as back-up. The problem, besides the high additional cost ($1,500-$2,500 for septic tank and leaching lines) is that the amount of leaching area required, based on average water consumption of more than 100 gallons per person per day, restricts applications to generally flat areas with soils that percolate well. The total reliance on septic tanks and leaching as an on-site waste management device tends to concentrate home development in areas that should be reserved for agricultural use.

There is no easy answer to the code dilemma facing the person who wants to build a composting toilet. As is true of building codes, health codes give discretion to the local health officer (usually a medical doctor) to accept alternative systems. But, again, few local officials choose to use the authority they have. Proprietary and generic types of composting toilets and greywater systems are now being tested in Oregon, California, and elsewhere. One approach is to request an experimental permit and agree to submit to regular laboratory tests that measure the fecal coliform present in greywater effluent or compost containing human fecal matter. If you are remodeling your existing house, you may not need to apply for a permit at all. Or you can call your concrete block compost privy a "root cellar."

This privy design has been submitted to many local and county health departments for approval, and experimental permits for construction have been received in some localities. Any waterless system represents a technical and psychological departure from the thinking and practice of many sanitarians. Their concern is to minimize individual involvement and responsibility for waste management. They tend to favor the "out of sight, out of mind" approach to waste management that centralizes the disposal of wastes in facilities run by public agencies. Systems which require maintenance by individual householders are discouraged because they create a control problem for local officials.

In the years I have been concerned with owner-building and ecologically sound living technology, I have seen that people do make a difference in getting local government to respond to the need for changes in building and health codes. The most striking example is in California where owner-builders have organized to get code reform for owner-builders in rural areas. Drought in California is sparking interest in waterless toilets and other conservation technology. A number of other states accept alternative systems on an experimental basis. The wheels move slowly, but only you and I can cause them to move at all.

Chapter 5.

Household Composting

The idea of composting your own waste is intimately tied to managing all wastes in your household. Together with the composting privy, these form an integrated approach to recyling all organic wastes. Notes on aerobic methods and design of composting bins were prepared by Bill and Helga Olkowski.

CHOOSING A LOCATION AND MAKING BINS

If possible, select a shady place so the piles will not dry out too quickly. The north side of the house or garage is often perfect. The ideal setup would be a ground area three or four feet wide and from nine to twelve feet long so at least three wooden bins can be constructed, each approximately a cubic yard in size.

The fronts of the bins should be made of removable boards, allowing easy entry as the contents are emptied. The sides and floor of the bins should fit as tightly as possible so bits of organic matter cannot fall through and provide overlooked fly breeding material underneath and outside the bins. Fit the lids tightly to keep out the rain. Keep one bin for storage of matter to be composted—general garden debris, but not manures or kitchen garbage—while you toss the going pile back and forth with a pitchfork between the two remaining bins.

You can expect your compost, made by the following method, to take about three weeks to complete.

Building the Pile
Start with some absorbent material on the bottom of the bin. Sawdust is good and you can easily obtain it from lumber yards, cabinet shops, high school woodworking classes. Then put down layers of green and dry matter and manure, if you are using it. If you use some other nitrogen source, sprinkle it over the layers as you go along. Make a three to five inch layer of each of your materials until the bin is full.

The smaller the size of the materials you put in, the more surface area you expose, and the faster the pile will decompose. For this reason you may wish to chop up coarse materials—melon rinds, dry weeds, stalks or straw—into shorter lengths (five to eight inches) with a cleaver.

When you have finished building the pile, you should have about a cubic yard of material. A smaller pile will not hold the heat adequately; a larger one is rather much to turn.

After you have built your pile, you may need to water it. It should be moist but not too wet, as there needs to be plenty of air throughout the pile. No liquid should run out the bottom. If this should happen at any time, put a thick layer of sawdust into the adjoining bin and turn the pile over into that bin to trap the juices.

Turning the Pile
After you have built the pile, let it sit a day or so. Then, with a pitchfork turn the compost into the neighboring bin, examining it while you do so. Turn the top, bottom and sides of the old pile into the center of the new bin, the center of the old around the edges of the new. Do this each time the pile is turned to ensure that all materials are exposed, eventually, to the heat of the center, killing any fly eggs, larvae, or plant pathogens.

Mix and turn your pile in this fashion at least every third day. Turning it more often (up to once a day) will speed the process of decomposition. If you have made your pile properly, for the first few days the internal temperature will rise, reaching approximately 160°F by the third or fourth day. It will return to this temperature for several days each time you turn the pile and then slowly cool. When it no longer warms up after turning, it is finished; it can be put out in the garden as soon as the temperature stays below 100°F.

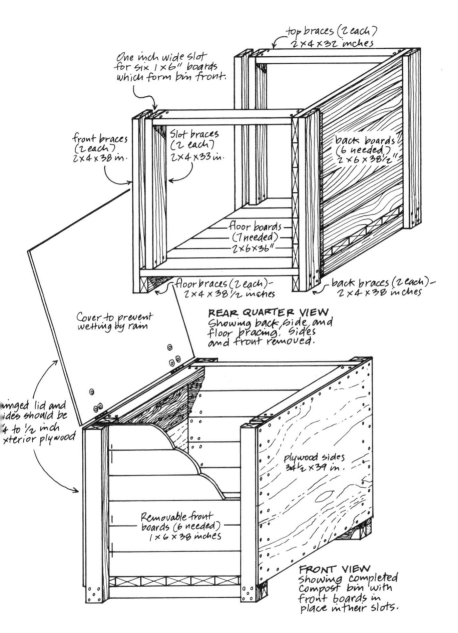

top braces (2 each)
2×4×32 inches

One inch wide slot
for six 1×6" boards
which form bin front.

front braces
(2 each)
2×4×38 in.

Slot braces
(2 each)
2×4×33 in.

back boards
(6 needed)
2×6×38½"

floor boards
(7 needed)
2×6×36"

floor braces (2 each)
2×4×38½ inches

back braces (2 each)
2×4×38 inches

Cover to prevent
wetting by rain

REAR QUARTER VIEW
Showing back, side, and
floor bracing. Sides
and front removed.

hinged lid and
sides should be
¼ to ½ inch
exterior plywood

plywood sides
34½×39 in.

Removable front
boards (6 needed)
1×6×38 inches

FRONT VIEW
Showing completed
Compost bin with
front boards in
place in their slots.

A homemade compost bin.

SUMMARY

Select and prepare composting area. Assemble materials. Chop as finely as possible, filling bin with alternate layers of green matter, dry or high carbon materials, and nitrogen source. At intervals while building pile, or when finished, add sprinkling of water. Turn compost every *second or third day* into neighboring bin using pitchfork. In turning, mix materials thoroughly—top goes to bottom, outside into center. Avoid spilling material outside bins. While turning, examine pile for fly larvae; center heat will kill them. Notice differences between piles as to age, particle size, moisture, odors: Does the pile smell like ammonia? Is it almost too hot to touch in the center? Is it moist? Finally, finish turning your compost pile with a flat shovel (and broom if necessary). Bin should be clean before introducing next batch.

Possible Problems

If compost does not heat up within two days: (a) You may need to add more nitrogen (blood meal, urine, etc.). (b) Your pile may be too dry. Add a sprinkle of water while turning. (c) Or, on the other hand, your pile may be too wet. Add sawdust while turning. (d) It may be you are not turning your heap often enough, and lack of oxygen is preventing warm up.

Compost heap is giving off strong ammonia smell: You have too much material high in nitrogen. Add a sprinkling of sawdust while turning.

Pile cools down but still contains some undecomposed material: Sift through coarse screen and return uncomposted materials to a new pile to go through the process again.

Using Your Compost

Compost may be used on top of beds around plants as a mulch. Finished compost may be spread out on newly harvested beds and turned under before replanting.

Use sifted finished compost in seed beds or flats. In carrot beds, for instance, a good mixture is one-third each sand, sifted dirt and sifted compost. Coarse particles can be returned to a new compost pile to break down further.

Unfinished (still hot) compost may be spaded into the ground *only* if it has had at least a week of composting and if no planting in that spot will follow immediately. Further decomposition of the material in the soil requires nitrogen, just as it does in the bin. Thus, raw compost may cause a temporary deficiency of nitrogen in the soil that will interfere with growing plants. Also, if the compost contains much manure that has not sufficiently decomposed, or if there is a strong ammonia smell coming from it, young plants may be damaged by the excessive nitrogen.

Caution: If ammonia smell persists heavily after the compost has cooled, do not use close to plants (within four or five inches of stems) as you risk causing nitrogen burn. Next time use proportionately less manure and more vegetable matter when making compost pile.

Chapter 6.

Greywater Systems

Good water is a real bargain—priced at a dime to a dollar per ton delivered at your faucet. Home water use averages about 150 gallons per person each day. About half of this is used for exterior watering, the other half is used indoors:

Daily Average Household Water Use, Per Person

Toilet	30 gal.	42%
Bath	20 gal.	32%
Kitchen	10 gal.	12%
Laundry	15 gal.	14%
	75 gal.	

Water use in the home can be reduced by changing fixtures *and* changing habits. During two years of drought in California many families found they could cut water use in half by reducing outside watering, flushing less frequently, taking showers instead of baths, and reusing greywater—that is, all dirty water except that from the toilet. Using a dry toilet saves almost a quarter of the pure water used in the house.

Another strategy is matching water quality to its end use. It makes no sense to use five to seven gallons of pure water to flush the toilet. Greywater can be used for flushing and for watering gardens. This chapter discusses how greywater can be used and various ways to collect, treat and recycle it.

Only recently has the distinction between sewage containing fecal matter and urine (blackwater) and greywater been made. Plumbing, building and health codes have yet to recognize the difference. Using greywater for above-ground irrigation around the home is new and regulations have not been adopted. Greywater systems should be designed to collect and treat greywater for above-ground irrigation in dry seasons, or to recyle it below the surface.

Non-toilet uses of water in the home amount to about two-thirds of total use, or about 30-50 gallons per person each day. Although greywater is free of fecal matter and urine, it may contain large amounts of organic material in the form of food particles, grease, suspended solids, phosphorus compounds and residues. The pollution potential of water is measured by the amount of oxygen required by bacteria to oxidize the waste materials in a given unit of wastewaster. The Biological Oxygen Demand (BOD) of household greywater may be as high as that of mixed sewage. Greywater *may* contain pathogenic organisms, but has a much lower probability of doing so than sewage. In any case, greywater in quantity must be treated before use.

The kind of treatment is related to volume of greywater produced, its sources, and intended reuse. The systems described in this section are divided into *low volume*—up to 50 gallons per day—and normal residential use of 100 gallons per day and more. Low volume systems will generally be limited to rural areas. Because of its high organic loading, greywater will quickly begin to decompose anaerobically, creating odor and attracting insects. In low flow situations it must be emptied every day, or, if stored, it must be filtered and disinfected. Normal flows require subsurface disposal in winter using a modified septic tank and disposal field; a diverter permits greywater use for irrigation in summer. Although the kitchen produces the least volume of greywater, it accounts for half the organic solids. Problems can be avoided by leaving the kitchen sink attached to the sewer or septic tank. Most greywater will require treatment through a combination of settling, filtering, and disinfection. Without treatment greywater can be used to irrigate orchards, vineyards, and seed, fiber, and pasture fodder crops.

SEPTIC TANKS AND SUBSURFACE DRAINFIELDS

Cesspools: In earlier years, the house waste line often emptied into a large masonry-lined and covered excavation—a cesspool.

Cesspools offer no pretreatment and tend to cause odors and to clog up; they are generally prohibited by codes.

Septic Tanks: The septic tank and subsurface drainage field are the standard way to treat and dispose of wastewater at the home site. One out of three dwellings is served by this system, which consists of a concrete holding and settling tank capable of holding three to five days of household wastewater (usually 1,000 gallons capacity for an individual home), and a network of perforated pipes laid in rock-lined trenches. The length of drainline depends on water use in the home and capacity of the soil to absorb effluent through the year.

The first design for a septic tank was patented in 1861. The operation of the system is biologically complex (although mechanically simple), including anaerobic fermentation, denitrification,

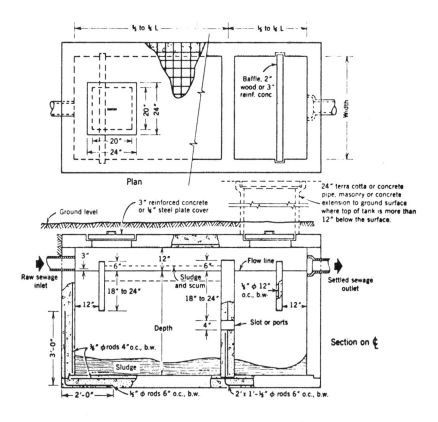

Details and dimensions of standard septic tank.

mouldering and aerobic composting. All household wastewater flows into the septic tank where its flow is slowed down by baffles. Sludges settle to the bottom where they are digested anaerobically. Scum and foam float to the top. The partially clarified effluent flows into discharge lines when the tank is nearly full.

Disposal fields offer another level of effluent treatment. Complex soil communities recycle nutrients, transform disease-causing organisms into harmless protoplasms, eat organic debris, reduce BOD, and evaporate water. Fields are designed to fit the absorption capacity of a given site. A trench must be dug, various layers of sand, gravel and crushed rock laid down, and pipe from the primary treatment tank installed and covered. There are a number of disposal field designs in existence which take into account the resting of trenches and the different absorption and treatment capacity of various types of soils.

Septic tank/drainage fields are water purifying and recycling devices. Areas in the southwest previously served by septic tanks discovered their ground water tables falling when they sewered up and all wastewater was discharged out of their area.

The use of the septic tank system is largely limited by the capacity of the soil to accept given quantities of wastewater. Where soils are poor or the water table high, it is sometimes possible to design above-ground mounds of sand and soil which disperse water by evaporation to the air as well as percolation through the soil.

Too often, blame for septic tank failure is placed on the system, while the cause often is simply lack of maintenance. The tank needs to have sludge pumped out every one to five years. Safe subsurface disposal is difficult in heavy clay soils, rock, or on slopes.

Reduction in water use will have a big impact on performance. If customary water consumption is cut in half, disposal fields would then be, essentially, twice as big.

Water saving devices installed in all California's subsurface-disposal served homes would save constructing a three-foot wide disposal trench twenty six thousand miles long, costing about one billion dollars.

Mechanical Aerobic Systems: Aerobic systems are small scale mechanical treatment units designed for home use in locations where site conditions do not permit use of the usual septic tank and disposal field. The principle of these systems is to force air through sewage to speed up decomposition and produce relatively clean effluents. In most aerobic systems, sludge must be removed every eight to twelve months. There are many such systems on the market, of varying

Typical section of disposal field trench.

design, cost and reliability. Two major issues confront the user: maintenance and subsurface disposal. The basic problem is that aerobic systems, unlike septic tanks, are mechanical systems with electrical pumps, filters, and aerators. They must be regularly and properly maintained. A second issue is that though it is often asserted that use of an aerobic unit reduces the length and increases the lifespan of a subsurface disposal field, research at the University of California and University of Wisconsin tends to refute these assertions.

Ozonation: Irradiation of wastewaters with ozone kills all microorganisms without the harmful residual effects of chlorine. Thus, ozonation is beginning to be used in large scale treatment processes. Some homesite treatment systems are now incorporating ozonation to provide additional treatment so that wastewaters may be safely used for irrigation.

Simple Greywater Systems: The simplest low flow system is the "Mexican drain"—a direct line between a greywater drainline and the garden. This requires modifying household plumbing as described later in this chapter. Connect a garden hose directly to a drain or collecting drum. This system can be used only in dry weather and requires moving the hose at the end of the line every few days.

If kitchen greywater is used, use a grease trap or rack filter to remove solids and prevent clogging of pipes and soils. A simple rack filter can be made out of a 55 gallon drum. Incoming greywater is

85

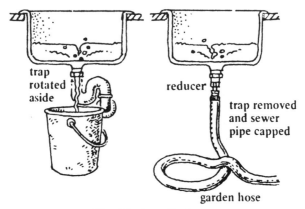

trap
rotated
aside

reducer

trap removed
and sewer
pipe capped

garden hose

The Mexican Drain.

strained through a rack containing a filtering material such as wood
shavings, or automotive or air conditioning filters. Clean the filters
every few weeks. Make sure to have a tight fitting top to prevent
insect breeding.

Small flow sand filters with immediate reuse can be constructed of
local, cheap materials. Boxes made of redwood, 55 gallon drums,
stone, masonry, or barrels can be used. This system must have a
hillslope gravity feed or a raised house in order to work. If below
ground level, greywater must first feed into a holding tank. It is then
pumped to the top of the sand filter and then either flows directly or is
pumped to the garden or pasture. In general, six to eight gallons of
greywater can be filtered by each cubic foot of sand per day.

A 55 gallon drum can safely filter 20 to 30 gallons each day. For
greater volumes, more drums can be added. Loading rates can be
greater in hot climates, with coarse sand, and with open filter boxes.
In rainy season, cover the sand filters as they will fail if continually
saturated. If the kitchen sink or lots of grease contribute to the
greywater, a grease trap or settling tank will prolong the sand filter's
life. Install a grease trap between your home and the sand filter. Sand
filters need occasional back-washing with clear water and removal of
the top one or two inches of sand.

Two small alternating sand filters will work longer than one big
filter because the resting filter gets a chance to aerate and dry out. Two
alternating filter beds will reduce maintenance. Each filter should be
able to receive an average daily flow. At the first signs of clogging,
switch greywater flow to the resting bed.

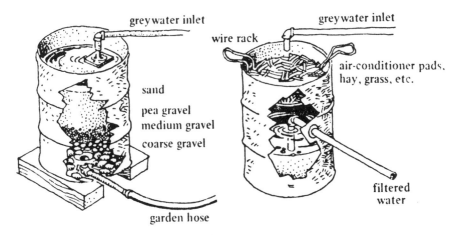

greywater inlet

greywater inlet

wire rack

air-conditioner pads, hay, grass, etc.

sand

pea gravel
medium gravel
coarse gravel

filtered water

garden hose

Rack filters.

Low Flow System Settling Tank: A simple settling tank for a low flow system has been designed by septic tank consultant T.H. Winneberger. Greywater flows from the house into a 30 gallon plastic garbage can set into the ground. Grease and floating particulate matter rise to the top. A plastic overflow pipe feeds into a check tank made from a five gallon plastic tank. If soils are not permeable the area around the settling and check tanks should be excavated and backfilled with rock and gravel, the finer sized gravel at bottom. When the tank is full, the effluent flows into a conventional drain field which should be sized about a foot in length for each gallon of estimated daily greywater flow. Build a frame and removable cover. Periodically, the settling tank will have to be cleaned of grease and scum, which can be composted or buried.

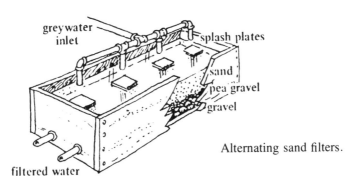

greywater inlet

splash plates

sand
pea gravel
gravel

filtered water

Alternating sand filters.

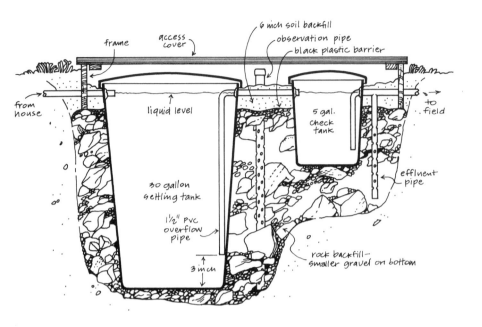

Low flow greywater tank (after Winneberger).

Greywater Treatment Tank: For normal household greywater flows, a scaled-down septic tank can be used together with a subsurface drainfield that may be reduced in length. For greywater flow of 150-200 gallons per day—within the range of an average household—a 300 gallon tank with 75-150 feet of leaching line should be satisfactory. Where it is desirable to use the greywater for irrigation in summer, a pumping chamber can be added to receive water directly from the septic tank and bypass the leaching field. This will require a simple diverter valve. The pumping chamber houses a submersible pump for irrigation.

This is an adaptable system meeting the needs of most households. The treated greywater should not be used for purposes other than irrigating nonedible plants. Using the septic tank for treatment purposes reduces maintenance, odor and insect problems. Less than four cubic feet of sludge and scum should build up annually, so the tank should need to be pumped clean only once every five years.

If higher quality greywater is desired, a sand filter can be added, either in combination with a septic tank or following a grease trap. A 1,500 gallon filter tank suitable for household use is illustrated.

88

Household scale commercial greywater systems are coming onto the market. These systems produce high quality effluent although the first cost is high—$1,000-$2,500 and more. Maintenance is unknown. The systems use a variety of straining, filtering and disinfectant techniques. The end product can be recycled for toilet flushing or irrigation.

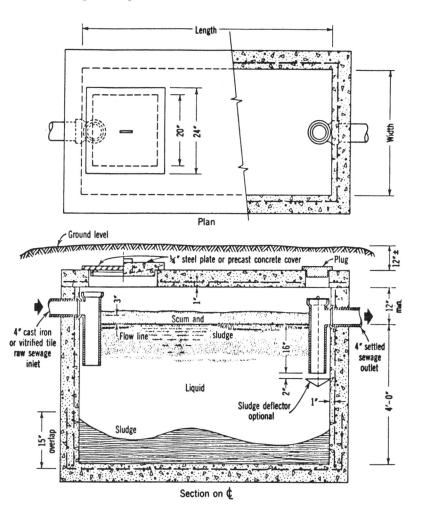

Small concrete septic tank suitable for holding greywater.

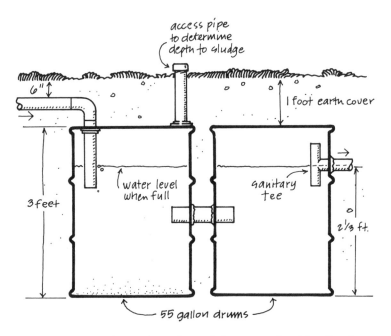

Labels in figure:
access pipe to determine depth to sludge

6"

1 foot earth cover

3 feet

water level when full

sanitary tee

2⅓ ft.

55 gallon drums

A greywater holding tank can be made by connecting 55-gallon drums in series.

MODIFYING HOUSEHOLD PLUMBING

Greywater can be separated and collected by modifying the standard household drainage system. This requires some knowledge of basic household plumbing. Don't attempt to replumb your system unless you are familiar with it. The basic idea is to add a separate greywater line before wastes flow into the main three or four inch pipe which services the toilets. The greywater line diverts water into a holding tank, or greywater treatment system. From there it can go into the garden in dry weather or into the leaching field or sewer during wet weather.

Greywater can be diverted selectively. Drainage from the kitchen sink is least desirable because it contains the highest concentration of

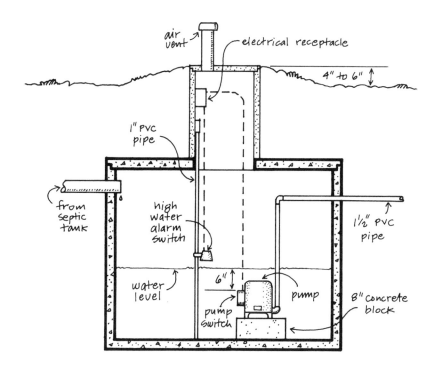

Section through greywater holding tank with pump for irrigation.

food particles, grease and soap, which can clog lines and cause odors. If you decide to use all greywater, then install thin mesh plastic screens at each receptacle to intercept food waste, hair, etc. If the kitchen sink is part of the system, the garbage disposal cannot be used.

Waste line modification should be made using ABS plastic pipe. (See diagram.) Install a double sanitary tee in the waste line; connect it to a threaded PVC ball or gate valve to allow diversion of greywater to a holding tank. When the valve is open, greywater flows into the storage tank; when closed, greywater flows as before into the sewer or septic system.

If your layout permits, the diversion line can empty into the greywater treatment system, or into a 55-gallon drum and from there

A large sand filter using modified septic tank.

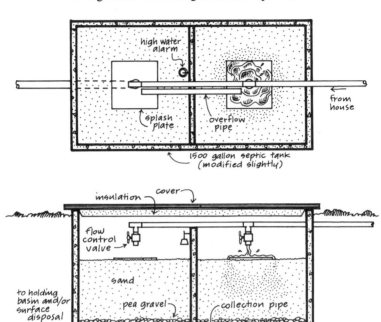

by gravity flow directly into the garden. If this cannot be done, and the collection point is below the garden, you may install a sump pump in a drum and pump from there.

Whether yours is a gravity flow or sump pump installation, waste water is best conveyed from tank to garden by ¾ " or 1 " hose: the

Greywater Sources and Quality: Average Home Daily Use

Order of Preferred Use	Gal/Day	Total Percent of Solids Produced
Bath/shower	80	27%
Bathroom Sink	10	27%
Washing Machine Utility Sink	35	23%
Kitchen Sink	25	50%
	150	

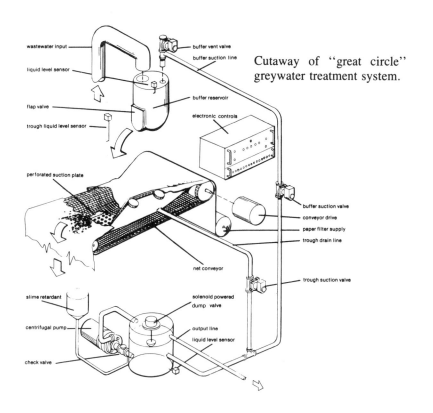

wastewater input

buffer vent valve

buffer suction line

Cutaway of "great circle" greywater treatment system.

liquid level sensor

buffer reservoir

flap valve

electronic controls

trough liquid level sensor

perforated suction plate

buffer suction valve

conveyor drive

paper filter supply

trough drain line

net conveyor

trough suction valve

slime retardant

solenoid powered
dump valve

centrifugal pump

output line

liquid level sensor

check valve

Schematic of Aquasaver greywater system.

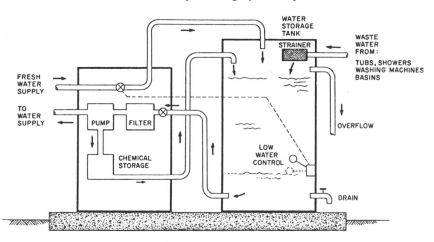

WATER
STORAGE
TANK

STRAINER

WASTE
WATER
FROM:

TUBS, SHOWERS
WASHING MACHINES
BASINS

FRESH
WATER
SUPPLY

TO
WATER
SUPPLY

PUMP   FILTER

OVERFLOW

CHEMICAL
STORAGE

LOW
WATER
CONTROL

DRAIN

bigger the better. A central hose may feed several lateral hoses connected by way of "wye" junctions so that waste water is distributed over a large expanse of garden or lawn. Rotate the arms around the garden frequently to reduce the possibility of localized puddling or excessive residue build-up.

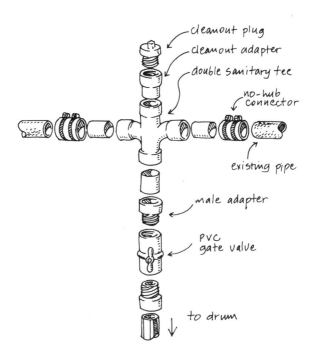

You can divert household greywater by installing "T" connection before toilet waste line.

Filtration of water may be achieved by attaching a cloth bag to the end of each lateral hose. The bag will serve to disperse the water overflow, intercepting particulate substances and soap residue. Remove the bags periodically, wash them and turn them inside out, sun dry, and re-use.

An essential objective of your greywater recycling system is to provide maximum conformance to health regulations and minimal disturbance of your home's existing sanitary system. The following design elements will insure that yours is a safe and sanitary system.

a. Do not alter or in any way modify the 3-inch and 4-inch lines; they carry the toilet wastes.
b. Direct drainage of all existing 1½-inch and 2-inch water lines into the sewer must be maintained. The diversion lines are installed as an *auxiliary system* to the conventional sanitary system, not as a replacement.
c. To insure gravity flow into the buffer or sump tank, the 1½-inch or 2-inch horizontal pipes must be installed to slope at a rate of ¾ inch per running foot.
d. Cleanouts should be installed at the entrance and exit ports of the buffer drums and at all intersections of drain pipes to allow maintenance of the waste lines.
e. Do not collect waste water in open tanks or reservoirs where children may gain access to the water or mosquitos may breed. The buffer and/or sump tank must have a tightly fitting lid.
f. Buffer tanks must have an accommodation for overflow into the sewer.
g. Do not allow any cross connections between supply lines and waste line modification.

USING GREYWATER IN THE GARDEN

Several principles govern the best usage of the water for garden irrigation. They are:

a. Apply waste water directly to the soil surface; do not overhead sprinkle or allow the recycled water to splash off the soil and contact the above-ground portion of plants. Greywater is not suitable for drip irrigation as the solid matter it conveys would clog the emitters in the pipe.
b. Distribute the greywater to flat garden areas; avoid steep slopes where run-off may be a problem.
c. Disperse the waste water over a broad area; avoid concentrating it on a single site.
d. When available, use fresh water for garden irrigation on a rotating basis with greywater to aid in leaching the soil of contaminants.
e. Apply thick compost mulches to areas receiving greywater to facilitate natural decomposition of waste residues.
f. Greywater is alkaline. Do not use it on acid-loving plants such as rhododendrons, azaleas, or citrus fruits. Lawns and deciduous fruit trees may receive greywater in rotation with fresh water.

g. If you must use waste water for irrigating food plants, restrict its application to the soil around plants of which only the above-ground part is eaten, such as corn, tomatoes or broccoli. *Do not* apply it to the soil around leafy vegetables or root crops. Best of all, use greywater for ornamental foliage and use what fresh water is available for your vegetable garden.

h. And, finally, use the waste water on well-established plants. Seedlings and houseplants will not tolerate the impurities in household waste water.

POSSIBLE PROBLEMS

What about soaps and detergents? Are they harmful to the soil and plants?

As a general rule, soaps are less harmful than detergents, but sustained use of greywater containing either one presents potential problems. The common problem of soaps and detergents is that they both contain sodium, an element which in excessive amounts is harmful to soils (destroying soil aggregation) as well as to plants (inducing tissue burn). The best strategy is to minimize the use of cleaning materials, and wherever possible choose soap rather than detergent. Gentle soaps, such as soap flakes, are preferred to those heavily laden with lanolin, perfumes, and other chemicals. Where detergents must be used, select those which do not advertise their "softening powers" (softeners are rich in sodium-based compounds). If you plan on reusing washing machine water, bleach should be minimized or eliminated, and boron-based (Borax) detergents absolutely avoided. Phosphates in detergents are not as great a problem in soil application as they are in sewage discharge into water bodies; nevertheless, low phosphate detergents are preferable. Ammonia is acceptable in reasonable amounts.

Must any precautions be taken to protect against damage to the soil with sustained use of greywater?

Over extended periods of greywater application, sodium may build up in the soil resulting in poor soil drainage and potential damage to plant tissue. High levels of sodium may be detected by conducting a pH test of the soil using litmus paper (obtained from a pharmacy or nursery). If the pH reading exceeds 7.5, the soil has become overloaded with sodium. Correct the problem by spreading gypsum (calcium sulfate) over the soil at a rate of two pounds per 100 square feet per month. Continue treatment until the soil pH drops to 7.0. As a

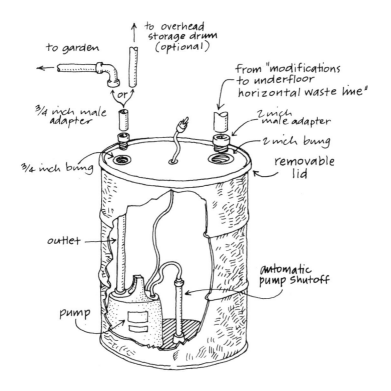

to garden

to overhead
storage drum
(optional)

from "modifications
to underfloor
horizontal waste line"

3/4 inch male
adapter

or

2 inch
male adapter

2 inch bung

3/4 inch bung

removable
lid

outlet

automatic
pump shutoff

pump

Greywater sump pump below garden level. Greywater is automatically pumped up to garden.

precaution against further sodium buildup, gypsum may be applied to the soil at a rate of three pounds *per month* for every 50 gallons *per day* discharge to the area being watered. Normal dilution of waste water by rainfall and/or fresh water irrigation will help to cleanse the soil of sodium. When available, use fresh water for garden irrigation on a rotating basis with greywater.

Is there any danger of pathogen transmittal by using greywater in the garden?

Waste water from the shower, bathtub and washing machine can conceivably contain disease-causing organisms. When the greywater is discharged into the soil, however, potentially harmful viruses and bacteria are quickly destroyed by the abundant soil organisms better suited to the soil environment. If the pathogens were to survive, it is unlikely in any case that they would be assimilated by the plant roots and translocated to the edible portion. Nevertheless, do not apply greywater to root crops which are eaten uncooked.

Chapter 7.

The Urban Sewer

Much of this book is concerned with how all of us can take more responsibility for our own shit to save water, soil and money. The 1970 census showed that out of 77 million dwellings, almost 20 million were served by on-site sewer systems—mostly septic tanks—while the rest were served by public sewers. If you have your own on-site system, read the chapters on dry toilets and greywater systems. They apply directly to your situation.

Most of us don't have that choice, but we can understand and possibly influence the kind and quality of wastewater treatment we pay our taxes for. Most city people don't know where their water comes from or, most especially, where their sewage goes. It's time to find out.

The cost of sewage treatment and collection will soon be the second greatest expense of local government. Only schools cost more. In the late 70s, the country is engaged in underwriting the largest public works program in history, larger even than the highway program of the 50s. The Federal Water Pollution Control Act of 1972 (Public Law 92-500), also known as the Clean Water Act, was part of the package of legislation growing out of the environmental movement of the late 60s and early 70s. The law provides grants and loans to local agencies throughout the country to pay for the planning and construction of projects that clean up waterways. Standards have been set for water quality and sewage discharges, which were supposed to be met by July 1977. The law states that "the best practicable control technologies" be used to clean up the water. In practice it hasn't worked that way. "Best technology" has in most cases been interpreted by engineers and administrators as the same old technological fix of flush toilets, sewers, central treatment, and disposal of treated effluent into the very lakes, rivers, bays, streams and oceans we are trying to clean up. Engineers and contractors have a firm grip on the program. Since consulting engineers are paid a percentage of the construction cost for designing a system, the more it costs, the more we pay, and the more they make. There is no incentive in this scheme to design low cost, ecologically sound systems. Ecologically based treatment, while using simpler engineering techniques, is more complex biologically—something "sewer rats" aren't comfortable with.

Most of these enormous sums is going into building huge regional systems of pipes and interconnectors with single large treatment plants. Builders and bureaucrats seem to favor these super sewers even though much of the money is spent simply moving wastewater around, rather than treating it. This means not only high construction costs but also high operating costs, since water is heavy and it takes enormous electrical energy to pump it. Single large treatment plants also tend to put all the eggs into one basket. Plant breakdowns can be disastrous; moreoever, the movement of so much water from its natural course creates ecological imbalances.

In smaller communities the problem seems more severe because people are closer to it. Until recently EPA favored centralized solutions over the development of better on-site technologies. The cost of constructing and operating central treatment systems—and only 12 percent of the construction cost is paid directly locally— often means either the near bankruptcy of local government or vastly increased taxes. As a result there is an increasing interest in the development of workable smaller scale systems. The investment nationally in present on-site systems is around $20 billion, by no means an insignificant figure.

All evidence points to the fact that clean water can never be achieved economically, or perhaps at any price, with today's technological fix. The cost is too high, the environmental and social disruption too great, the health problems connected with discharge are serious and growing. Potable water is no longer limitless. The future lies in smaller systems, less transport, low or no water systems, reclamation rather than discharge of wastewater, land application of nutrients, and full biological treatment through conversion of nutrients into edible plants and animals.

Where water transport sewerage exists, there are ways to reclaim the water and nutrients. Standard treatment plants have been favored in a society where the main value of land is as an economic commodity rather than as an irreplaceable asset. The main criterion for treatment has been to get rid of the stuff as fast as possible. Mechanical aeration reduces retention time to a matter of hours, so more sewage can be moved out in less space. Alternate ways of purifying wastewater require more time and more space.

THE ANATOMY OF SEWAGE TREATMENT

Primary sewage treatment involves three steps: screening, then grinding incoming sewage to remove debris and other non-decompostable substances, followed by detention in large settling tanks, not unlike septic tanks where solids settle out on the bottom as sludge and anaerobic digestion takes place. Methane produced by digestion is used to heat the settling tanks. Primary treatment removes about forty percent of the pollutants through gravity settling and anaerobic digestion.

Secondary treatment, through oxidation, began to be adopted in the U.S. starting in the 50s. This treatment is essentially a mechanical compression in time and space of what happens naturally in free flowing waters. The effluent remaining after settling is pumped into aeration tanks where large volumes of compressed air are blown in from the bottom and sides, keeping the water well supplied with oxygen, nature's greatest purifying agent. The water is also kept in a constant state of agitation to help absorb outside air. Air bubbling through the tank allows aerobic bacteria to multiply thousands upon thousands of times.

In addition to the effluent and air, a third ingredient is added to the aeration tank: sludge from a recent batch which has already made a cycle through the aeration tank. This "activated sludge" picks up suspended solids in the waste water and adds to the supply of aerobic bacteria. Sewage stays in the aeration tanks about five hours, then is emptied into final settling tanks and chlorinated before release into waterways. About 95 percent of solid organic wastes are removed in

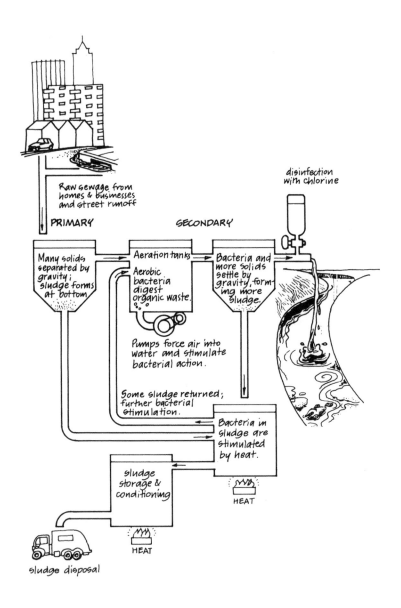

Diagram of standard secondary sewage treatment.

the process, although a large percentage of the nutrients nitrogen and phosphorous, as well as insoluble compounds, remain to produce pollution.

The stabilized sludge remaining must be disposed of. Common means are landfill, burial, burning, or, in some cases, drying for use as fertilizer. Heavy metals from urban and industrial wastes may be present in wastewater and, if so, these toxins may preclude the use of sludge for fertilizer.

ALTERNATIVES TO CENTRALIZED SEWAGE TREATMENT

There are alternatives to the mechanical and concrete jungle approach to sewage treatment and disposal. The simplest approaches use natural purifying processes found in healthy communities of soil and water organisms.

Oxidation ponds are shallow man-made ponds or lagoons. Sewage is purified through the natural action of sun, wind, bacteria, algae, snails and other detritivores feeding on decomposing wastes; sludge settles to the bottom where it decomposes anaerobically. Oxidation

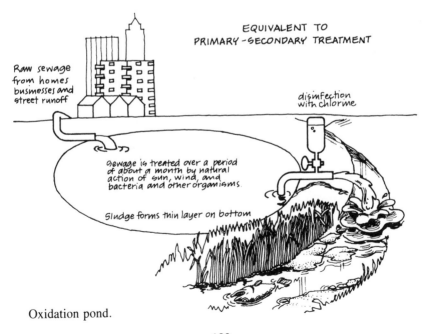

EQUIVALENT TO
PRIMARY-SECONDARY TREATMENT

Raw sewage
from homes
businesses and
street runoff

disinfection
with chlorine

Sewage is treated over a period
of about a month by natural
action of sun, wind, and
bacteria and other organisms

Sludge forms thin layer on bottom

Oxidation pond.

ponds are widely used in small communities and for special purposes such as the digestion of feed lot wastes. Sewage needs to be retained a month or so to provide the equivalent of secondary treatment. Construction and operating costs are low, although a major drawback is the large amount of land required.

A variation on the lagoon system combines purification of wastewater with the application of both reclaimed nutrients and recycled water on agricultural or recreational lands. After a time in the treatment lagoon, often with some mechanical aeration, nutrient-rich effluent is applied to croplands by spray irrigation. Contact with the air and with microorganisms in the soil and plants purifies the water so the end product passing through this living biological filter will be safe to drink. The sludge settling on lagoon bottoms is periodically dredged up and applied to adjacent soil as a conditioner. Crude variants of this type of treatment have been used for centuries. So-called sewage farms go back to the 15th century in Europe. Systems developed in Berlin and Paris in the late 1800s filtered wastes naturally through sandy soil. Compared to present alternatives, the land required by the early sewage farms was about one acre per thousand people. Aerated oxidation ponds require one acre for 5,000 people; while a trickling filter-activated sludge secondary treatment requires only one acre per 50,000 people.

Conventional treatment processes cannot remove or detoxify the majority of the most harmful components of modern day wastewater such as pesticides, herbicides, phenol and a host of other complex chemicals that cause cancer and otherwise endanger health. Although conventional systems are biological in the sense that they use bacteria to oxidize wastes, research has shown that such monoculture systems are less effective and more unstable than polyculture systems using a variety of species, including bacteria, invertebrates, detritivores, algae and plants. All these are usually present in the normal lagoon, which accounts for its success in producing clean water. For example, a lagoon system at St. Helena, California, removes 99.9 percent of the biological oxygen demand (BOD) and over 90 percent of the nitrogen and phosphorous. Similar results can be achieved through high technology mechanical systems, but only at many times the cost, including the use of massive amounts of electricity and chemicals. Water hyacinths have proven to be a particularly effective natural cleanser of wastewater. Research by NASA demonstrated that one acre of water hyacinths in a pond can remove over 3,500

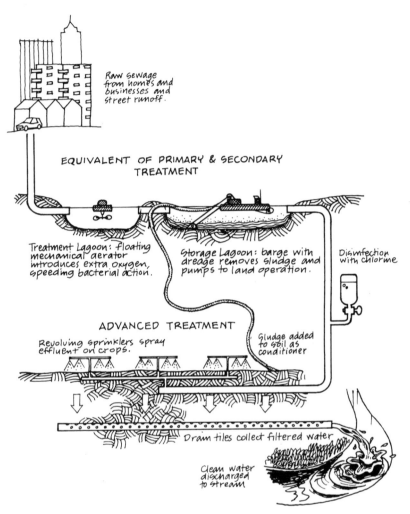

Raw sewage from homes and businesses and street runoff.

EQUIVALENT OF PRIMARY & SECONDARY TREATMENT

Treatment Lagoon: floating mechanical aerator introduces extra oxygen, speeding bacterial action.

Storage Lagoon: barge with dredge removes sludge and pumps to land operation.

Disinfection with chlorine

ADVANCED TREATMENT

Revolving sprinklers spray effluent on crops.

Sludge added to soil as conditioner

Drain tiles collect filtered water

Clean water discharged to stream

Aerated lagoons and spray irrigation: a natural water treatment system.

pounds of nitrogen, 800 pounds of phosphorous, 18,000 pounds of toxic phenol and absorb 44,000 grams of heavy metals. Lagoon systems, however, require a lot of land to treat sewage, and climatic changes affect their performance and reliability.

A recent advanced variation of full biological treatment has been developed experimentally by a group of San Diego aquaculturists.

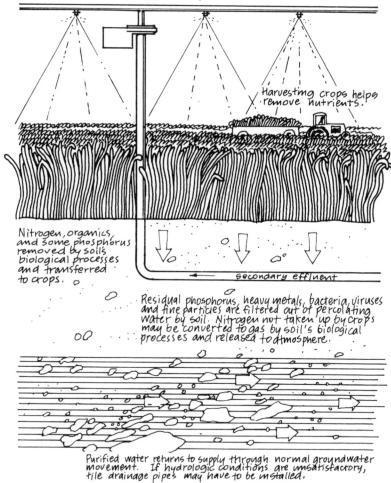

Slowly revolving spray-irrigation rig spreads effluent on soil and crops.

Harvesting crops helps remove nutrients.

Nitrogen, organics, and some phosphorus removed by soil's biological processes and transferred to crops.

Secondary effluent

Residual phosphorus, heavy metals, bacteria, viruses and fine particles are filtered out of percolating water by soil. Nitrogen not taken up by crops may be converted to gas by soil's biological processes and released to atmosphere.

Purified water returns to supply through normal groundwater movement. If hydrologic conditions are unsatisfactory, tile drainage pipes may have to be installed.

Detail of spray irrigation system. Water is purified by passing through the soil filter.

The *Solar Aquacell System* is a composite of the best design aspects of four well proven technologies: 1) Aerated lagoons for wastewater treatment; 2) A culture of floating aquatic plants—water hyacinths—to remove nutrients and toxins on wastewater; 3) Polyculture of microinvertebrates, fish and shellfish for maximum removal and concentration of nutrients and organics in wastewater; and 4) Solar heated greenhouses for environmental control.

The system consists of rectangular ponds covered with greenhouses to retain heat, artificial habitats to increase the biologically active surface area in ponds, channels to control water movement and flow, aeration systems and sand filtration for maximum water quality. As raw wastewater flows through the system of modular ponds, or *aquacells,* nutrients and toxins are taken up by the root structure of rapidly growing hyacinths aided by small aquatic creatures which reside in the root structures on plastic gratings or biograss. Water hyacinths from the first cells most

Water hyacinths growing on raw sewage at Solar Aquafarm's pilot plant near San Diego.

contaminated with toxins are used for fuel in a methane digester that provides heat and power for the operation. Partially purified wastewater leaving the first cell is ozonated for safety. Subsequent cells are home for fish and giant freshwater prawns. The prawns are grown in individual cubicle-like cages submerged in the pond. My trip to the experimental plant was like a visit to a beautiful pond: there was no smell, just the gentle gurgling of water passing through the variegated colors and textures of hyacinths and duckweed, complemented by the flashing gold of beautiful Japanese koi.

Root structure of the water hyacinth showing fine root hairs which soak up nutrients and provide a home for aquatic critters.

A mature specimen of the freshwater prawn grown in the Solar Aquacell plant.

Since treatment plants are smaller in scale, the aquacells can be located closer to the raw material; less money and energy is required to transport sewage. Construction and operating costs are half that of conventional secondary sewage treatment. Energy costs are lower. Although more area is required than for conventional treatment plants, the aquacell plant can treat wastes of 10,000 people per acre—a land use one-tenth that required for open lagoons. Natural biochemical processes remove viruses, pathogens, and toxins, even those virtually impossible to remove through conventional sewage treatment. The aquacell system creates a bounty by providing new local jobs and a year round harvest of fish, shrimp and lovely aquatic plants.

By trading expensive concrete, steel, chemicals and electricity for natural processes using earthen ponds, greenhouses, solar energy and hardy pollution-consuming plants and invertebrates, this process promises to demonstrate that raw wastewater can be converted into high quality reclaimed water with a bonus of useful by-products, at one-half to one-quarter the treatment expense of conventional methods.

All systems employing natural biological filtering make a great deal of sense. They are economical to build and operate, don't require much fossil fuel or electricity, are non-polluting (since no effluent is discharged into waterways) and place reclaimed water and nutrients on the land where they are needed. Why, then, do we still see relatively few purification and reclamation projects built on these principles?

One reason is that as our major urban areas were built up, land was not set aside to act as a cheap natural recycler. Of course, there are significant exceptions. The brilliant and single-minded creator of San Francisco's Golden Gate Park, John McLaren, built that park out of a sandy wasteland by diverting sewage from San Francisco's main outfall. The British Garden Plan enthusiasts of the early twentieth century recognized the importance of creating an ecological niche—a greenbelt around cities to absorb wastes—but their plans were seldom carried out.

Of equal significance, waste has always been the province of sanitary engineers, who have favored complex engineering solutions over simple biological solutions.

Engineers in the environmental health field have set extremely stringent, if not unreasonable, standards for zero-discharge systems, even though they are infinitely safer and less polluting than conventional systems. So we continue to pay for the high economic and environmental costs of conventional treatment plants.

We have been making technological choices that have been displacing products and processes which fit in with the cycles of nature. Then to rescue nature, we have been applying "environmental technology" which substitutes for natural processes, and therefore duplicates the work available from the ecological sector. This displacement and duplication is a crippling economic handicap.

Consider the example of how we handle human waste. As the growth of urban areas has become more and more concentrated, much energy, including research and development work, has gone into developing and implementing technologies to protect our lakes and rivers and coastal waters from the wastes we are dumping into them. These wastes, however, are themselves rich sources of chemical energy capable of being recycled back to the farmlands from which these nutrients came. They would replace much of the fertilizer that we produce from fossil fuels and eliminate the need for energy-expensive tertiary sewage treatment.

—Sen. Mark Hatfield in *Energy: Today's
Choices, Tomorrow's Opportunity.*
(World Future Society,
Washington, D.C.)

Conversion of wastewater to valuable by-products through aquaculture.

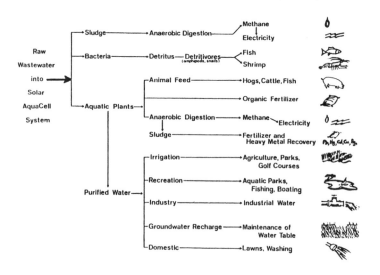

What may be our largest agricultural waste is not usually recognized as such, but is thought to be both an urban product and an urban problem: the tons of garbage and sewage that are burned or buried or flushed into rivers. This, like all waste, is the abuse of a resource. It was ecological stupidity of exactly this kind that destroyed Rome. The chemist Liebig wrote that "the sewers of the immense metropolis engulfed in the course of centuries the prosperity of Roman peasants." The Roman Campagna would no longer yield the means of feeding her population; these same sewers devoured the wealth of Sicily, Sardinia and the fertile land of the coast of Africa.

—Wendell Berry in *A Continuous Harmony*
(Harcourt Brace Jovanovich)

When shit has value, the poor will be born without assholes.

—Old Portuguese saying

HEALTH AND THE LAW

When confronted with alternatives to flush toilets and sewers, many sanitary engineers and health officials are apt to say, "We have a system that works. Why try something else?" Many go on to credit water-based systems with the virtual elimination of intestinal diseases—cholera, typhoid, typhus and dysentery—that plagued urban dwellers in Europe and the United States before the discovery of the biological basis for disease by Pasteur and others.

Ironically, these intestinal diseases and many others are spread primarily through water. Water transport and disposal of body wastes gives the bacteria feet. Only through using enormous quantities of germicides (chlorine and more recently ozone) in the treatment of sewage effluent and drinking water have we been able to keep the nasty bugs under control. One system now being proposed uses gamma radiation from nuclear wastes to sterilize water.

Since its early days discharged sewage has often been the main culprit in the transmission of fecal-borne disease. With the first widespread use of water closets, death rates rose. From 1859 to 1900, cholera epidemics in London and other cities were traceable to contamination and leaking of sewage into water supplies. Often the rich who could afford water closets were victims of cross connections which, unknown to them, leaked sewage into their sinks. In rural areas, where privys were common, epidemics were less severe.

Garbage collection, basic sanitation, and safe water supply all came into being at the same time sewers did. It was largely protection of water supplies from fecal contamination—the careful separation of water and waste systems—that led to elimination of fecal-borne disease, as well as widespread use of chlorine which is a powerful antibiotic agent.

When sewers came into being during the late nineteenth century, some medical authorities were critical. One writes:

> Sanitation is a purely agricultural and biological question; it is not an engineering question and it is not a chemical question, and the more of engineering and chemistry we apply to sanitation, the more difficult the purifying agriculture. Our houses are flushed away but we pay for it by fouling every natural source of pure water. If there comes an outbreak of typhoid as often as not we find the drains to blame; but as a matter of fact we prescribe more drains as a remedy. We doubtless manufactured typhoid retail in the old days, but with the invention of the water closet we unconsciously embarked in the wholesale business.

> —Dr. Vivian Poore, M.D., quoted in
> *Conservancy or Dry Sanitation versus*
> *Water Carriage,* J. Donkin (1906)

All too often expertise sacrifices the insight of common sense in its preoccupation with technique. The "Manual of Instruction for Sewage Treatment Plant Operators," published by the New York State Department of Health, accurately and perhaps unconsciously points to the absurdity of the technology it is instructing:

> Sewage consists of water plus solids which are dissolved or carried in suspension in the water. The solids are very small in amount, usually less than 0.1 percent in weight, but they are part of the sewage that presents the major problems in its adequate treatment and disposal. The water provides only the volume and a vehicle for the transportation of the solids.

Progress toward finding better ways to reclaim wastes without dilution with water or discharge to waterways is often slowed because of health risk questions. Yet the problems with water based disposal continue to multiply:

Within the last several years, research and various surveys have shown that dangerous viruses such as hepatitis and polio are not killed by chlorination in central treatment plants, and are able to survive for months in discharge waters which are used for drinking or recreation downstream; that about half the sewage plants in the country do not operate properly on occasion because toxins in industrial waste kill

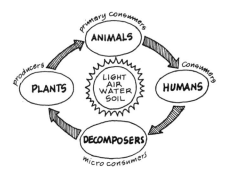

The nutrient cycle.

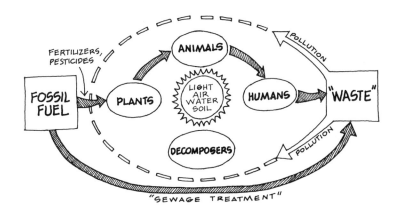

The cycle broken.

bacteria essential to the process; that chlorine, the major disinfectant used in sewage treatment, combines with other chemical compounds commonly found in wastewater to produce cancer causing agents and chloroform in drinking water; that chlorine in effluent dumped into the ocean combines with salts to form toxic acids; that the chloroform created by chlorine is released into the atmosphere and is destroying the earth's protective ozone layer; that large scale deposits of sludge dumped many miles out at sea are creeping back onto beaches, making them a ghastly unusable mess; that sludge disposal on land is also becoming a major problem; that spray produced by aeration processes in modern sewage treatment plants contains viruses and other pathogens which contaminate surrounding areas; and that, currently, the discharge of secondarily treated sewage into fresh waters is estimated to use up all available oxygen in the water, turning the fresh water into anaerobic sewers in spite of cosmetic "treatment."

Clearly, the health issue is not one of absolute safety, which is, in any case, unprovable for any technology; the issue is *acceptability*. Health and safety laws arrive at inferred levels of acceptable risk based on cultural values and economics as well as scientific evidence. For example, a whole body of law specifies automobile safety, yet each year in California alone over 4,500 people die using this technology and more than 300,000 are injured. We accept the risk and continue to reject proposals that could lower it.

The acceptance by health and sanitation people of better ways to manage "wastes" depends upon persistent and informed pressure.

In most places, methods of disposing of household wastewater must be approved by local health and building departments and a permit taken out for their construction. Generally, the only approved method is the standard septic tank and subsurface disposal field. To get a permit for a septic tank you must submit percolation tests that indicate soil conditions can handle the projected flow. Health officers have the authority to approve other types of systems, yet the dry toilets and greywater systems described in this book are not generally accepted.

So, if you wish legally to install one of these systems, you may run into resistance from local authorities. Here's why: 1) As a general rule bureaucracies like to deal with other bureaucracies rather than with people. The more a problem can be centralized, the better. 2) It is easier to come down on the little guy than the big guy. I have heard of cases where permits were denied until the applicant could prove that dry toilets killed viruses, which no conventional sewage treatment can do. 3) Regulatory processes are set up for routine. It is easier to say "no" rather than to rethink the problem and design a better procedure. 4) Like most bureaucracies, the assumption of regulation follows Murphy's law: "If anything can go wrong, it will." Alternate systems *do* require more individual responsibility than conventional systems. Bureaucracy assumes you are incapable and unwilling to take responsibility for such basics as managing your own waste. 5) Regulations are oriented towards control, not towards education. Most people who are using or have built alternate systems are valuable sources of information for neighbors. Yet the person who rocks the boat is often viewed as a part of the problem, rather than part of the solution. 6) A keystone of sanitation practice is water borne sewage. For years the United States Census measure of progress was the number of flush toilets in the country. Any departure from this practice, no matter how rational or safe, is looked at as a step backwards. 7) Most local health departments are not set up to evaluate unconventional systems.

A number of states are evaluating alternatives and pressure is on for federal government to take a look at alternate systems. In the meantime, if you plan to apply for a permit: 1) Be prepared to present test data providing positive soil percolation. Authorities may require proof of backup capability for a conventional system. 2) Show an understanding of basic system operation and existing data (this book should help you). 3) Prepare an outline of maintenance procedures. 4) Offer to present data from an approved laboratory testing for fecal coliform, salmonella and parasite ova.

115

EPILOGUE

Any technology divorced from the whole flow of nature tends to produce a condition that poet Robert Graves calls "mechanarchy": the perfection of technological means to produce a chaotic sterile environment. The current technology of "waste disposal" (the term reveals the syndrome) is still fighting a war against nature, built on fragments of nineteenth century science not yet integrated into an understanding of life processes as a unified, but cyclical, whole.

The discovery in 1828 by the German chemist Wohler that urea—a basic plant nutrient—could be chemically synthesized, led to the still current belief that naturally occurring nutrients in organic matter could be considered "waste," since they could be readily manufactured.

The discovery some years later that then-common diseases were caused by microscopic germs led to chemical weapons such as chlorine to effectively kill microorganisms and thus shorten the natural cleansing process occurring in soil and aquatic communities. The resulting reduction in enteric disease was coupled with a vast increase in pollution by water-based sewage. In spite of the public health and environmental hazards of central sewerage most people continue to believe we have a "clean machine."

The invention of efficient means to convert the stored solar energy of coal and oil to mechanical work created the belief that the energy flow of natural systems was not essential to our well-being. We are just beginning to realize that a society which is tied to fossil fuel cannot last very long.

The challenge of appropriate technology is to design a high quality environment for people that makes use of the innate harmony and productivity of natural systems. Doing this means freeing ourselves from destructive technological fixes and finding ways to create a high quality of life without destroying the life sustaining balance of nature.

Images and possibilities come to mind.

Flying low over the flat expanse of California's great central valley, my eyes follow the thin white line of the California Aqueduct. Clear waters from a thousand tumbling Sierra streams unravel their way southward in a concrete ditch, pausing at the edge of the Tehachapi where giant pumps push the flow over the mountains into a vast artificial lake: Lake Parris, a giant flush tank for millions of Los Angeles toilets.

Flying back we skim over the Santa Clara Valley—a wide flat ribbon running south from San Francisco Bay between two lines of rounded coastal hills. Thirty years ago the most productive orchard lands on the continent, this once lush valley floor now grows an unbroken mat of roofs and streets—a ticky-tacky crop of asphalt and air pollution made possible by sewer lines and freeways running down the center of the valley.

The sewage plant stands on a ruined marsh next to the dump, its aerating booms slowly pivoting around the center of circular concrete ponds. I wonder when the marsh, a "sewage plant" of unseen complexity and intricate beauty, will once again take up its delicate dance of life and decay while the concrete monument slowly moulders to its more basic molecular form.

In the windowless conference room on the twelfth floor of the Health Department building, a large group argues the health risks of compost privys. At the coffee break I pull a gallon jar from its brown paper sack under my chair and gently sprinkle the sweet-smelling stuff aged in our privy on the plants in the room. I imagine the plants smiling and smile back.

In not too many years, our vast cities and settlement patterns will be very different from now. To survive the waning days of the oil age, cities will have to reassemble themselves into coherent biological systems.

Cities with millions of people living on coastal desert plains using water pumped in from six hundred miles away hang on a very slender thread. As the city devoted to the care and feeding of the automobile fades, streets will be torn up and gardens planted. The soil, now compressed and lifeless dirt, will be restored to life with our composted wastes and greywater. Like the hill towns of Italy which for centuries perched themselves on the rocky unproductive hills, reserving the rich bottom lands for food, the pattern can be reversed so that the ruined agricultural valleys can bloom again, and the hills will be terraced with gardens and houses. In the cities, wind-powered solar heated aquacultural greenhouses will grow fish and shrimp on wastewaters and return purified water for use in the home. The soft edges of wetlands and marshes, cushions against flood and superb biological filters of impurities, can be restored.

The shapeless and disintegrating urban mass bound together with cars, clocks, cheap fuels, TV and institutionalized waste can be recreated into many communities, each with its own history, its own limits, and its own future.

Annotated Bibliography

I. HISTORY

The Bathroom. Alexander Kira, Viking, 1976, Penguin, 1976, Bantam, 1977.

An overview of the culture and technology of hygiene with suggestions for design improvements of fixtures. Ironically, although the book is presented as an objective study, the author reveals his own hangups: there is no mention of what happens to wastes when they leave the bathroom, no discussion of water use, and the photographs of people using bath and showers (in bathing suits) have their eyes blacked out!

Conservancy or Dry Sanitation versus Water Carriage. J. Donkin, 1906.

A good example of the books from the period that criticizes the biological absurdity of water transport systems then being constructed. Too bad no one listened.

Clean and Decent: The Fascinating History of the Bathroom & the WC. Lawrence Wright, University of Toronto Press, 1967.

A delightful and well illustrated account of how Europe muddled through the last 2,000 years.

Dirt: A Social History as Seen Through the Uses and Abuses of Dirt.
Terence McLaughlin, Stein and Day, 1971.
The whole scoop on dirt and filth in merrie, dirtye Olde England. A coprophiliac's delight with fascinating details—it's a wonder anyone survived into the twentieth century.

II. CENTRAL SEWERAGE AND BIOLOGICAL TREATMENT

New York State puts out an excellent series of manuals designed to teach technicians who run water treatment and sewage treatment plants. The manuals give a good introduction to the chemistry and microbiology of water and sewage, explain testing procedures as well as the basic processes and hardware connected with centralized systems. Gives you an inside view, clearly written and illustrated.
Manual of Instruction for Sewage Treatment Plant Operators.
Manual of Instruction for Water Treatment Plant Operators.
The Design of Small Water Systems.
Laboratory Procedures for Wastewater Treatment Plant Operators.
All available from: Health Education Service, P.O. Box 7283, Albany, New York 12224.

Clean Water, Leonard A. Stevens, Dutton, 1974.
A modern critique of the big sewer engineering mentality with excellent descriptions of reclamation land treatment alternatives.

III. DRY TOILETS, COMPOSTING AND WATER SAVING

Composting. Clarence Golueke, Rodale Press, 1972.
"The Bible" by a most respected authority.

Residential Water Conservation. Murray Milne, California Water Resources Center, UC Davis, March 1976.
A thorough review of the state-of-the-art in water conservation devices, the economy and psychology of water savings. Descriptions and illustrations.

Stop the Five Gallon Flush! Witold Rybczyoski & Alvara Ortega, id. School of Architecture, McGill University, Montreal, rev. ed., 1976.

The first survey of alternative waste disposal systems and dry toilets growing out of the ecological architecture movement. Covers all systems but little detail.

Composting: Sanitary Disposal and Reclamation of Organic Wastes. World Health Organization, Geneva, 1956.

The classic technical source on composting: the numbers and hard facts here on composition of compostible materials, aerobic and anaerobic processes, techniques, hardware, and sanitary management of composting processes and products.

Excreta Disposal for Rural Areas and Small Communities. E.G. Wagner, J.N. Lanoix, World Health Organization, Geneva, 1958.

Another WHO monograph that exhaustively covers design, construction and operation of privys, aqua-privys, latrines, cesspools, septic tanks—all manner of low-tech village scale sanitation.

Wastewater Treatment Systems for Rural Communities. Commission on Rural Water, Washington, D.C., 1973.

This book is a nicely designed guide to techniques and components available for managing wastewater (including sewage), short of the central treatment plant. An appendix (about half the book) surveys existing types of equipment suitable for home or village use. However, no evaluation is made. It's a handy guide of what's around and how it works. Some review of basic theory.

Village Technology Handbook. VITA (Volunteers for International Technical Assistance) College Campus, Schenectady, New York, rev. ed., 1970.

VITA produces simple designs for village development in poor countries. The handbook includes easy to follow plans and designs for elementary water and waste systems.

Goodbye to the Flush Toilet. Ed. by Carol H. Stoner, Rodale Press, 1977.

The respected Rodale organization has been active in promoting alternatives through their conferences, publications and research support. This book is a comprehensive review of composting toilets and greywater systems with a historical overview.

IV. GREYWATER AND ON-SITE DISPOSAL SYSTEMS

Septic Tank Practices. Peter Warshall, Doubleday, 1978.

A modest title for a book that clearly lays out practical aspects of various types of on-site sewage treatment, and the relationship of treatment to soil, water use, construction, maintenance and politics. Written by a brilliant biologist who has integrated theory with a practical hands-on approach.

Cleaning Up the Water: Private Sewage Disposal in Maine. Maine Dept. of Environmental Protection, July 1974.

A good pamphlet on acceptable alternatives by a state that has pioneered better ways to handle waste on-site.

Manual of Greywater Treatment Practices. J.H. Winneberger, Ann Arbor Science, 1976.

V. HEALTH

Environmental Sanitation. Joseph A. Salvato, John Wiley and Son, 1958.

The standard text in the field, covering control of communicable diseases, water supply, sewage and waste treatment, public health standards and enforcement in housing, food handling, insects and vermin. Many sanitarians learned from this book. A good presentation of accepted practice (which by the way includes privys): "If properly located, constructed and maintained they can be adequate, economical, and sanitary devices for the disposal of human excreta."

Away with All Pests: An English Surgeon in People's China, 1954-1969. Joshua S. Horn, Monthly Review Press, 1971.

A fascinating account of the grass roots approach to good health through preventive measures. Good account of China's waste management.

The Toilet Papers was set in Times Roman
by Charlene McAdams; photographs by Wayne McCall;
camera work by Santa Barbara Photoengraving;
printed by Princeton Academic Press.